Understanding Stoves

For Environment and Humanity

Sai Bhaskar N. Reddy

UNDERSTANDING STOVES

STOVES

For Environment and Humanity

Sai Bhaskar N. Reddy

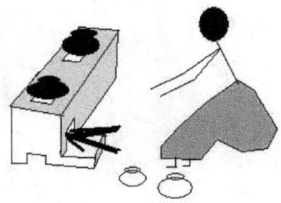

Understanding Stoves

For Environment and Humanity

1st Edition, released 2012

(OK) This is an 'Open Knowledge' book as declared by the author.
Dr. N. Sai Bhaskar Reddy, 2012
saibhaskarnakka@gmail.com
http://goodstove.com

Front cover photo: A tribal woman cooking on a three-stone stove.
Back cover photo: Author creating awareness on good stoves in a village.

Note: The author drew sketches and mostly took the photos presented in this book.

Published in 2012 by

MetaMeta

Paardskerkhofweg 14
5223 AJ 's-Hertogenbosch
The Netherlands
www.metameta.nl

For 'The Earth'

Foreword

This book presents the results of a decade of dedicated promotion and experimentation with fuel efficient stoves in India by my friend Sai Bhaskar Reddy Nakka. Sai Bhaskar's approach has been original in many ways. He has investigated the different mixes of biomass used in different areas of India – not just fuel wood but also leaves, dung, pellets, briquettes, seeds and other materials. He then designed the different stoves accordingly. Sai Bhaskar also looked at different products from the stoves – obviously the direct heating but also the possibility of producing charcoal (Biochar) as a by-product from the stoves. This work is not lab-based but undertaken in the promotion campaigns he has been involved within Andhra Pradesh and parts of other states in India.

The relevance of this work is large. Many people are and will remain dependent on local biomass for their daily needs. This issue of biomass has often associated with the degradation of the surrounding environment: the depletion of natural tree stands or the loss of nutrients to the growing cycle. It has also been associated with hardship: collecting biomass – typically by women – over large areas, and health issues because of excessive smoke from in-house fires. Because of such reasons the constant improvement of household stoves and stoves used in institutions and others is as much required as developing new methods of energy production – be it from algae, solar, the earth's heat, water currents or winds – or safe methods to extract fossil fuel.

Moreover, there are important developments to make biomass consumption more and more renewable. In several areas I have seen farmers starting to grow fuel wood on a commercial base, either on field boundaries or as special plantations. We recently came across this in Yemen where farmers had set up acacia plantations with the main aim to produce charcoal.

For all these reasons I am very happy that this book of 'The Stove Man' could be prepared, to share the praxis-based knowledge developed by a unique innovator and to serve as a source of inspiration and new ideas for those working in energy and development. We hope that this book gives an impulse to the constant effort to make lives better and richer.

Frank van Steenbergen

Table of Contents

Picture 1 A school Girl cooking food for the family

Introduction

This book draws from my seven-year journey towards understanding stoves and the people using them. It gathers my experiences, observations and insights into stoves. This work has given me immense happiness, for I believe it can help the cause of environment and humanity. I was involved in designing stoves[i], improving existing stoves, testing stoves and conducting studies. During the course, I interacted with communities and worked with them: creating awareness, sensitivity and on capacity development.

Picture 2 Smoke from stoves is a hazard for women

According to World Health Organisation (WHO, 2011)[ii], around 3 billion people cook and heat their homes using open fires and leaky stoves that burn biomass (about 2.7 billion use

On Understanding stoves; Authors journey into designing stoves; the value of stoves and the importance of facilitation of stoves.

wood, animal dung and crop waste) and coal. Nearly 2 million people die prematurely from illnesses related to indoor air pollution from household solid fuel use. Nearly 50 % of pneumonia deaths among children under five are due to particulate matter inhaled from indoor air pollution. More than 1 million people a year die from chronic obstructive respiratory disease (COPD) that develops due to exposure to such indoor air pollution. Both women and men exposed to heavy indoor smoke are 2 - 3 times more likely to develop COPD. Given this, it is estimated that there is a need for about 1 billion good stoves across the globe.

In India alone, about 570,000 premature deaths among poor women and children every year, and over 4 % of India's estimated greenhouse emissions are attributed to the impacts of cook stoves[iii].

Understanding stoves

Improving stoves is an effective way of improving the environment and serving communities. My work in this area provided me a great opportunity to meet a diversity of people and visit many places. Some of the primary stakeholders I could meet were the poorest of the poor: tribal communities living in remote areas, rural

Picture 3 People living on pavements, cook food on the road. Location: Mumbai City, India

communities, people living in urban slums, sex workers, artisans etc. Besides, I also met representatives of various government / non-governmental organizations, policy makers and scientists.

I had never imagined that I would be exposed to such a mosaic of human beings and understand so much about the environment through stoves. My work got the press' attention. They called me- *'Poyla* Sir' (Stoves Sir), *'Gramala Nestham'* (Friend of villages) and *'Manchi Manishi'* ('Good Man'). I consider these no less than honors bestowed upon me.

Over time, my work on stoves expanded into other research area – biochar / biocharculture, sustainable agriculture, safe water, habitats, sanitation, etc. Through the process, I came to realize the importance of 'Open Knowledge'[1] to achieving the common good. Therefore, I have

FREEDOM

OPEN KNOWLEDGE

declared all my work so far as open and freely available to all who may need it or find it interesting.

Adoption of good stoves by communities is an effective means to mitigate harmful environmental impacts. It also addresses climate change-related issues, since it is linked to the reduction of greenhouse gases (especially Carbon Dioxide) and the reduction of black carbon, an important factor accelerating global warming. Black Carbon accelerates global warming through excessive heat absorption, browning of skies (especially as observed in the south Asian region) and reduced albedo from the sky.

Improved stoves can help carbon sequestration, such as when the biochar[iv] from them is applied to the soil to improve its fertility.

---

[1]   Open Knowledge: The belief and commitment that knowledge and technologies relevant to common people should be open and not be patented for the common good. http://okgeo.org

Other advantages of improved stoves include conservation, sustainable use of biomass, and reduced indoor air pollution.

In most countries, biomass is still the main source of fuel for many people. Energy is always a concern, as a resource that is scarce, diminishing, costly, time-consuming, uncertain in supply and often inaccessible.

Poor people spend a considerable amount of time on sourcing the biomass for fuel. The urban poor visit backyards of industries / railway lines to look for fuel. Sometimes security personnel barter sexual favours with women, in exchange for letting them access these areas. There are instances of women being raped when they go wandering around common lands or forest areas looking for fuel wood. When there is no free access, people are forced to buy biomass, sometimes at very high prices.

There is a need for large-scale awareness and education regarding the extent to which access to fuel can affect the quality of life, especially that of the less-privileged.

There is no value to an innovation or technology relevant to common people if it is held up in patents

My journey into stoves

One day I was in a village[2] trying to understand the rural energy situation in India. A distinct observation was that in the courtyards of many houses, there was heaps of fuel wood. This fuel wood was collected, for use in cooking for several months. It was harvested from the village commons and / or the nearby-degraded scrubland. On

Picture 4 Fuel wood stored in the courtyard of houses

an average, people spend up to 4 days a month collecting fuel wood. Considering the rural wage rates being currently offered under government programs, it amounts to no less than $ 10. A majority of the rural poor earn around $1 per day. Their spending on fuel wood sometimes works out costlier than the Liquid Petroleum Gas (LPG) promoted by the government through a range of subsidies[3].

---

[2]   Srirangapur Village, Kondurg Mandal, Mahabubnagar District, Andhra Pradesh, India.

[3]   Fuel is highly subsidized in India; Government of India is bearing a cost of Rs. 30.10 per liter on kerosene and Rs. 439.50 per 14.2-kg domestic LPG. (March 2012).

I have visited about 50 household kitchens. I noticed that about half the stoves were 'three-stone' stoves and the rest made from clay. Such stoves are highly inefficient, rudimentary and primitive. I will never forget those visits, for they made me realize what a large number of people are stuck with them and exposed to their harmful effects.

The stoves and the kitchens do not reflect the changes in the lives of the families over the years. Many people are now living in concrete houses, eating higher-value food, can afford mobile phones, can send their children to English-medium schools, have satellite television connections, motorbikes, tractors to plough fields... and so on.. Despite all the winds of change and development, the kitchen remains a smoky place with dark-sooty walls and roof. Even traditionally, kitchens have occupied relatively small spaces inside or outside the house.

I decided that my first step would be creating awareness about indoor air pollution and its harmful effects. I went to communities and showed them pictures of their stoves through a projector. Although people in general were interested in watching the slide show, women objected when it came to seeing their own stove. This is because the kitchen was not a place of pride for them, even though they respect the stove very much. I then explained to the communities the impact of indoor air pollution and its multiplier effects on them. They were quite interested in adopting efficient stoves, but were helpless, as they did not have access to them. In one of the villages, the state government had promoted chimney stoves with grates about 15 years back. Those stoves were nowhere to be found. Within a few years, the stoves had disintegrated, the chimneys had been choked with soot, the grates burnt down and the stoves disappeared. The design of the stoves did not take into account how adaptable they were to local conditions and practices. The stoves had been distributed under a government scheme and were subsidized. Nevertheless, they could not find acceptance or sustainable demand from the local community.

This exposure made me sensitive towards the issue of stoves in rural areas[v] and I was motivated to help improve their stoves' design. I realized that there could not be a single globally good stove, and that the challenge was adapting designs to fit local needs.

As I started off, I could not easily access specialists in the field of good stoves. However, through the internet, I found numerous, diverse stove designs that had evolved around the globe - in response to a variety of food habits, cooking methods, cultural traditions, types of available biomass fuel, family sizes etc. Studying them formed the basis of my initial understanding of stoves, and subsequent research and design efforts.

Biomass stoves for sanctity

Many people say that biomass cook stoves should be banned altogether. However, all the elements of sanctity and sanitation can be found in them: Fire, Smoke, Charcoal and Ash. Cooking on biomass stoves is also a way of engaging with and respecting the elements of nature. Fire is highly respected even by the most primitive of communities. Fire is considered auspicious by many spiritual beliefs / systems / religions / societies. Some of them even worship it as a deity. In many religions, praying rooms are attached with kitchens. Considering smoke to be harmful does not fit with the culture of using incense sticks for worship, common to many religions.

Picture 5 Fire, Smoke, Charcoal and Ash - sanctify a place

The stove also signifies the cultural connection of the community with fire. As mentioned above, fire is widely regarded as a deity and biomass stoves adorn kitchens also used as prayer rooms. It is the medium through which offerings are made to the Gods as a plea for common good such as rains, good harvest, peace and harmony. In India, this is evident in ceremonies like *Yagnas*, *Homas* and *Agnihotras* (or Holy Altar). Cooking on biomass stoves, thus, represents this cultural respect for fire.

In rural areas, where poor people use fuel wood to cook, there is always some smoke. Smoke from biomass stoves protects them from a variety of pests. In many countries, termites are a major cause of economic losses. In many rural communities, seeds collected for sowing in the next season are commonly stored in the kitchen. Smoke protects thatched huts from pests such as termites and woodborers. It is very common in such places for people to use smoke as a means to repel mosquitoes and prevent diseases like malaria. On the other hand, cockroaches and insects are common in kitchens using Liquefied Petroleum Gas.

Picture 6   Seed is stored in kitchen as the smoke protects and preserves it.

Picture 7   Unattended thatched house roof, attacked by Termites

Exposure to smoke also increases the longevity of thatched structures. It has been observed in tribal habitats that, if unattended, the roof of a hut could be destroyed by termites within 3 to 6 months. On their part, thatched roofs are porous enough to allow smoke and soot to exit the hut.

Charcoal and ash are two other important by-products of the biomass stove. Many poor communities use charcoal for cleaning teeth, and the ash for cleaning utensils. Ash from holy altars is used as a vermillion to apply on the forehead and other body parts. At times in small quantities, it is also recommended as medicine by faith healers. In agriculture, the charcoal is used as biochar for improving soil-health and fertility. Ash is an important additive for acidic soils and an important source of potash and phosphate.

Some smoke is good and too much smoke is bad

Facilitating good stoves

Human beings have stepped on the moon, and explored the solar system and planets many light years away. However, we still have not been able to provide efficient cooking systems for millions of people living on this Earth. Providing one billion efficient stoves is a mission as important as any other mission is.

Picture 8 Creating awareness to children on good stoves

Yet, this mission is yet to be achieved. There are not many professionals trained in the 'Good Stoves' technology. We require many of them: researchers, designers and trainers. Universities and Institutes should conduct professional courses to meet this challenge.

As a policy, only a few countries have taken up the cause seriously. Crucially, the lack of awareness means that the issue has not caught the imagination of the youth. And as for the media, the silent death of millions due to indoor air pollution does not seem to have enough news value. Given the sheer number of people who are affected, it should be their responsibility to understand the issue and spread awareness about it.

It is very rare to find professionals with relevant education and training working on stoves. During awareness programs conducted in the schools, I would often ask the children one question: "There is an important task on earth yet to be accomplished i.e., facilitating one billion good stoves. How many of you will be willing to work on this in the future?" Rarely did anyone put his or her hand up. A common refrain was that 'our parents will not approve of it'. Besides, it was all too noticeable that even the word 'stove' was not mentioned anywhere in the school curricula. This reflects the general lack of emphasis on the subject, which also explains the lack of initiatives on the part of the government.

Thanks to the global reach of e-groups and networks related to stoves, there is now scope for interested stakeholders to come together, share, understand, design and facilitate good stoves. Some of the most important international networks include 'Improved Biomass Cook Stoves' (website and e-group)[vi], 'Partnership for Clean Indoor Air' (PCIA)[vii], 'Household Energy Network' (HEDON)[viii], 'Global Alliance for Clean Cookstoves'[ix], 'Winrock'[x], 'Deutsche Gesellschaft für Internationale Zusammenarbeit' (GIZ)[xi], ARECOP[xii], Aprovecho Research Centre[xiii], et cetera.

Whatever progress the human beings are doing on this earth, we can take is no pride in them if people are still using three-stone stoves

Picture 9 Participation of women in choosing Good Stoves

Biomass fuel for stoves

A variety of biomass is used as fuel for cooking-stoves. Biomass can be sourced from various types of plants, and in many forms - trunks, branches, sticks, twigs, bark etc. Regionally, some types of biomass are preferable to others, because of cost, access, calorific value, density of wood, ease of fuel wood preparation, whether they are fast growing species, easy to dry, etc. Exotic, invasive species such as *Prosopis Juliflora* and

Picture 10 Head load of fuel wood being carried, an arduous task

> 🗌   On biomass as fuel for cooking; conditions of cooking and the use of fuel.

Acacia plants are highly preferred as fuel wood in many tropical countries.   Some of the indigenous forest species come under federal / provincial forest legislations, which means that it is illegal to use them for this purpose (among others).

  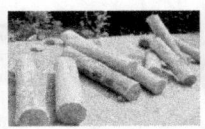 

Picture 11 Types of biomass used as fuel

Crop residue is another important source of fuel, available seasonally. Sometimes the loose biomass is converted into briquettes or pellets for use in stoves. In some stoves, loose biomass is used in various forms such as sawdust, wood shavings, rice husk, leaves, grasses, seeds, corncobs, shells (groundnut, coffee, etc.), etc. In places where biomass is not available, animal dung is also used as fuel. Dung / clay is used with charcoal dust for making briquettes / balls for use in stoves. Preferably, naturally occurring biomass or biomass available as by- product of various grades should be used. It is easier to come up with a good design for a stove if one knows what fuel would be used in it.

Picture 12 Briquettes and wood shavings used for cooking

Wood moisture is another factor important to efficient combustion. Wood moisture varies seasonally, based on the sunshine, precipitation (rain, snow, fog, etc.). In many tropical and semi-arid environments, wood moisture varies between 12 to 18 % on a dry day.

Picture 13 Wood shavings and saw dust used as fuel

Thin wood dries easily and has greater surface area as compared to large-sized wood. Very thin wood (less than one cm in diameter) is used for kindling. Sometimes, kerosene is used (about 5 to 10 ml) to start the fire.

Papers, cardboard, wax, camphor, vegetable oil, pinewood, pinecones, seeds etc., are also used for starting the fire. Communities for cooking (except for kindling) do not prefer thin wood. Thin wood burns conveniently with very less smoke, but sometimes emits excess soot. The stove users do not like soot deposit on their utensils and their kitchen walls. Similarly, bark of plants, although good for fire, sometimes leads to excess soot and smoke. The kerosene added for starting the fire can also emit quite some soot.

Torrefaction (French for roasting) of biomass (e.g. wood) can be described as a mild form of pyrolysis at temperatures typically ranging between 200 to 320 degree centigrade. During torrefaction, the biomass' properties are changed in a way that, as a fuel, it leads to much better combustion and gasification. It turns the biomass into a dry product, leaving no scope for undesirable biological activities like rotting. Torrefaction of biomass in combination with densification (pelleting / briquetting), is a promising step towards achieving logistical economy in large-scale sustainable energy solutions. Being lighter, drier and stable in storage, such pellets / briquettes are easier to transport, store and use.

Often, thin and neatly cut sticks are used for testing the stoves for efficiency and performance. However, the wood used in domestic cooking stoves is about 5 to 7 centimeters in diameter. The wood is never split perfectly into square cross sections, say 2 x 2 centimeters, like the wood pieces used for testing stoves. Splitting the wood in that manner is very difficult and demands lots of energy and time too.

Picture 14 Fuel longs of more than 6 feet length used in stoves

Typically, the length of the pieces of prepared / collected fuel wood is about 40 to 100 centimeters. Carrying, bundling and transporting them is not convenient if they are much shorter. Across India, it is common for people to carry fuel wood by balancing it on their heads or on bicycles. In tribal areas, people can be found carrying whole trunks and branches of trees on their heads, without cutting them into small and thin sticks. These are often 150 centimeters to 350 centimeters in length and nearly 10 centimeters to 30 centimeters in diameter. It is also common to use whole trunks in stoves, especially among communities living in/around forests.

Facilitating Good Stoves is simple to achieve.... but we are far away from doing that

Cooking conditions

In semi-arid regions, cooking is often done in semi-ventilated conditions, as rainy days are few. Cooking pots are covered with lids while cooking. Sometimes, while stirring / simmering of rice etc., the lid is removed completely or covers the pot only partially. While making *Rotis* (flat Indian bread), the pan is completely uncovered.

Picture 15 Food being cooked with half covered lid

Poor families often do not have separate kitchen areas at all. Even in well-off households, kitchens are often dark and dingy. Ventilation is poor, windows/ ventilators are rare.

Even in well-ventilated kitchens, traditional stoves cause a lot of indoor pollution. No one enjoys cooking because, for many, cooking amounts to smoking. Sometimes, such stoves emit as much smoke as 50 to 100 cigarettes, over 2 to 3 hours. Cooking during the rainy days often leads to highest release of smoke, due to higher levels of moisture in the wood and the stoves.

Cooking is mostly done early in the mornings (or the domestic rush hour – when children are leaving for their schools and adults are leaving for work) and in the evenings (when the children and the adults return). During these two periods, cooking becomes all the more difficult. Exposure to smoke from inefficient stoves makes cooking even more difficult. Those who cook, therefore (and it is mostly women), do so with little willingness.

Picture 16 Cutting trees for wood is one of the decentralized form of depleting biomass

Fuel feeding

Cooking on stoves that need to be fed fuel wood continuously, is like feeding a baby. It requires immense patience. One has to start the fire with the thinner sticks and then move to the ones with larger diameters. It is important that the ignition be done at the tips. Sometimes, users push the sticks too deep into the stoves. The stove then bellows out smoke the way a baby throws up when overfed. Therefore, using a given type of stove efficiently requires a certain amount of experience with it. 'Novice' users should be careful not to judge a stove's efficiency too quickly. Good stoves are designed in a way that allows for proper air circulation and mixing inside them. This makes them easy-to-use even for the less-experienced users.

Certain types of stoves, such as most of Top Lit Up Draft (TLUD) stoves, allow for 'batch-loading' of fuel. This means that the biomass fuel (chips of wood, wood shavings, small bunch of sticks, pellets, etc.) can be loaded just once, and then ignited at the top. As the pyrolysis[4] moves down the heap, wood gas is released which mixes with the secondary air just underneath the pot and combusts. After a batch of fuel is exhausted completely, charcoal / biochar[xiv] gets accumulated at the bottom of the combustion chamber. It can then be emptied and reloaded.

---

[4] Pyrolysis is a thermochemical decomposition of organic material at elevated temperatures with very little participation of oxygen. It involves the simultaneous change of chemical composition and physical phase, and is irreversible.

Batch-loading fuel stoves are not convenient for communities to adopt, as one needs considerable experience to handle them. For example, there is always a chance that one could add too much or too little fuel. Despite controls to regulate natural draft and forced air, controlling the power of the stove could get very difficult indeed.

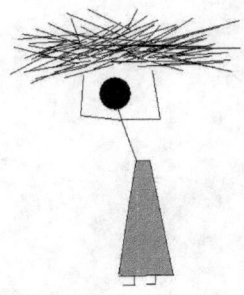

Picture 17 A woman cooking on traditional stove

Physical aspects of stove

Size of stove

A stove of 8 inches height is ideal. Any stove too high (about 12 inches or more) is inconvenient. Women prefer sitting comfortably on the floor while cooking. In rural areas, many women carry out many daily activities while sitting. Especially when making *rotis* (flat bread), they prefer a low stove.

Picture 18Women prefer to sit while cooking

Scientifically, the '*chimney effect*' in tall stoves would manage the smoke and aid combustion more efficiently. However, as per the ergonomics and local practices, they are not convenient to the user.

Often the pan for making *rotis* is about 10 to 12 inches in diameter. A 4 - 5-inch diameter stove, the stove needs to be operated at high power for the flames to spread underneath the pan sufficiently, for the *roti* to be completely cooked. Stoves

> ☐ On Physical aspects of the stove – size of the stove; types of material used for production of stoves and the efficiency of the stoves; and capacities of the domestic stoves.

around 6-7 inches in diameter are ideal, given the various types of the utensils used by an average-size family. Two of my stove designs were failures, for having a small diameter and for being too tall. The designer needs to strike a balance between the scientific principles and the user's requirements.

Sensitivity leads to awareness, knowledge and action. It is more important to be sensible than to be knowledgeable

Stove material and efficiency

Thermal efficiency is also dependent on the mass of the stove. In Water Boiling tests, relatively lightweight metal stoves prove to be thermally more efficient, especially during the cold start high power[5] phase. During the hot start high power[6] phase they show slight improvements in thermal efficiency. Heavy stoves, on the other hand, are thermally less efficient during the cold start phase, but due to retained heat

Picture 19 Proud owners of an efficient stove

and during the hot start phase they exhibit higher efficiency. In real-life conditions, high-power cooking is not preferable as the taste and the value of the food cooked is not as good compared to normal cooking. In addition, a compact stove is more adaptable than one with many movable or removable parts. Stoves made of two or more materials are costlier and sometimes less durable.

---

[5] Cold Start: Tester begins with the stove at room temperature
[6] Hot start: Tester conducts the test after the first test while stove is still hot.

Insulation and refractory quality are two important factors in stove-design. A material, which has both the properties, is of great value. The refractory material options currently available in the market are of high density and are bad insulators. Moreover, the cost is very high: each refractory brick costs about $1.

Firebricks or refractory bricks usually contain 30-40 % Aluminum oxide or Alumina, and 50 % silicon dioxide or silica. A stove made of them is guaranteed to last up to 10 years. However, its thermal efficiency is compromised. By combining lightweight insulating firebricks along with regular firebricks, thermal efficiency can be improved.

There is much difference between lightweight, insulating firebricks and heavy, dense firebricks. Insulating bricks are refractory and withstand very high temperatures, but their thermal conductivity is lower than required. They do not absorb the heat well at all. These bricks are mainly used for heat insulation. They are used on the outer side of the stove (around the heavy firebrick walls or under the floor bricks and the slab used as a foundation for the stove) to keep heat from the combustion chamber from escaping, and to thus achieve higher thermal efficiencies. Insulating Fire bricks are made up of Alumina: 37 %, Silica: 61 % and Ferric Oxide: 1.6 %.

The cost of the stove goes up when constructed using refractory bricks and insulating firebricks. Overall, the weight of the stove also increases, making it inconvenient to transport. While such stoves are more durable, their volume and surface area are relatively high and they lose a greater amount of heat through radiation (where only refractory bricks are used). The cylindrical and hemispherical shaped stoves have less surface area and radiate much less heat, as compared to box-shaped stoves.

Air is the best insulation media. Even by using the thin metal sheets for combustion chamber (with side air and secondary air holes), the life of the combustion chamber can be increased significantly. Even as air is sucked into the combustion zone, the combustion chamber is being continuously cooled. This increases the durability of the combustion chamber.

As a measure of safety from excessive heat emitted from the body of the stove, an external wall can be made out of a thin sheet or wire mesh. The surface area of contact between the hot parts and other parts of the stove should be minimum possible. This can be achieved by using pointed / thin screws, ceramic watchers, etc., thus minimizing the heat transfer between the grate / combustion chamber and the outer body. Many traditional stoves do not have such safety walls around them. The space between the inner and outer walls is often left vacant or filled with insulation material, such as broken pieces of glass, coarse sand, ash, pumice rock pieces, rock wool, etc., which are refractory and insulatory.

For transport and spread over wider areas, the weight of the stove should ideally be around 2 to 5 kilograms. Stoves weighing more than 10 kilograms are more difficult to market as a sustainable enterprise. They are also more difficult for a team of people to demonstrate and promote over a large area. Often, stoves promoted through subsidies /

Picture 20 An efficient good stove – Avan series, for cooking mid-day meals to the children

government programs are too heavy. Therefore, those in charge of promoting them do so unwillingly and ineffectively. The fact is that a majority of the stoves available in the market and sold on a large scale (although many are inefficient) are always lightweight. Being light, they cost less and are less durable. The lightweight stoves are much less likely to pass the simmering test[7] than the heavier stoves, if the stove user is novice.

Ideally, the holes in the stove grate should be small, but not so big as to affect its durability. Too much or too little primary air entering the stove from below, and through the grate into the combustion chamber, hampers the stove's performance.

---

[7] Simmering test: This portion of the test is designed to test the ability of the stove to shift into a low power phase following a high-power phase in order to simmer water for 45 minutes using a minimal amount of fuel.

There are stoves with scope only for primary air; they are also called blower stoves, because they are often found with blowers. It is common to see charcoal stoves often fitted with blowers underneath the grate. Grates made up of forged iron are cheap and durable. Many improved stoves fail because the

Picture 21 Air blower fitted to a stove, supplies additional air and creates draft for efficient combustion

grates fail. Grates need to be most durable of all the parts in a stove. In the *'Good Stove' Avan Series* design, the grate is avoided. Over a raised platform, fuel wood sticks are fed horizontally. The sticks do not fall into the combustion pit due to gravity, and the fuel wood is combusted only at the tips. Through the side hole at the bottom, the primary air reaches the fuel wood from below.

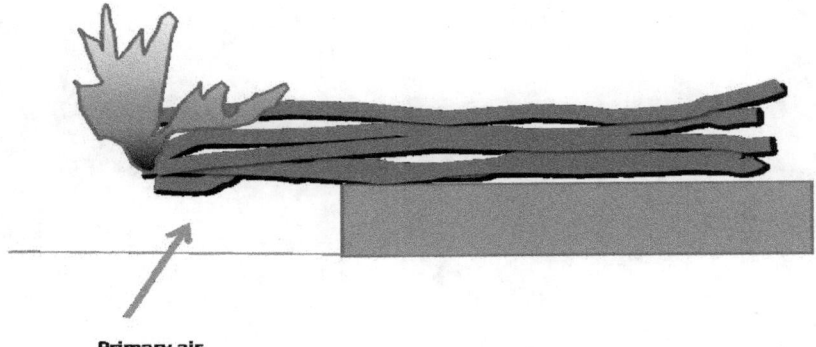

Primary air

Picture 22 Primary air comes from the wood, supports pyrolysis and combustion (Good Stove – Avan Series).

Capacity of domestic stoves

A domestic stove often serves, on an average, a five-member family. When guests visit, the same has to support, say, 10 members. It is very difficult for many families if their stove cannot provide this flexibility. On an average, cooking pots are 1 to 6 liters in capacity. Similarly, the number of people to be served through institutional stoves (used in big establishments such as restaurants / hostels / schools / religious places, etc.) is highly variable.

Life is being curious, knowing, acting and experiencing

Flame

Observing the flame is very important, and photographs are more useful than videos for the purpose. In a dark room, observing the flames gives a better indication of the stove's performance. One of the reasons is that a blue flame is more visible in the dark.

Picture 23 Flame from a wick lamp emits very good light.

☐   On types of flames; combustion process; colour of flame and behaviour of flame.

Combustion process

In the combustion process, the flame is the simplest thing to understand. Take a matchstick, light it and you will observe the following things: while striking a matchstick, due to friction and the chemical reaction, the flame starts at the tip. The matchstick head contains either Phosphorus / Phosphorus Sesquisulfide as the active ingredient, and gelatin as the binder. The process of combustion of wood (as in a matchstick) is like a controlled chain reaction. The initial heat (about 200 plus degrees) generated by the chemicals will pyrolyse the matchstick and release combustible gases. Usually, matchsticks are thin and made of lightwood, and so get pyrolysed easily at low temperatures. If the head of the matchstick is raised, only a limited amount of heat from the flame would reach the stick.

This would prevent pyrolysis and the matchstick might be extinguished. If the tip of the matchstick is lowered, the matchstick will pyrolyse rapidly and combustible gases would be released more. The flame will be large. Therefore, keeping the matchstick horizontal, or its head tilted slightly towards the ground, would help combustion.

Picture 24 A burning matchstick

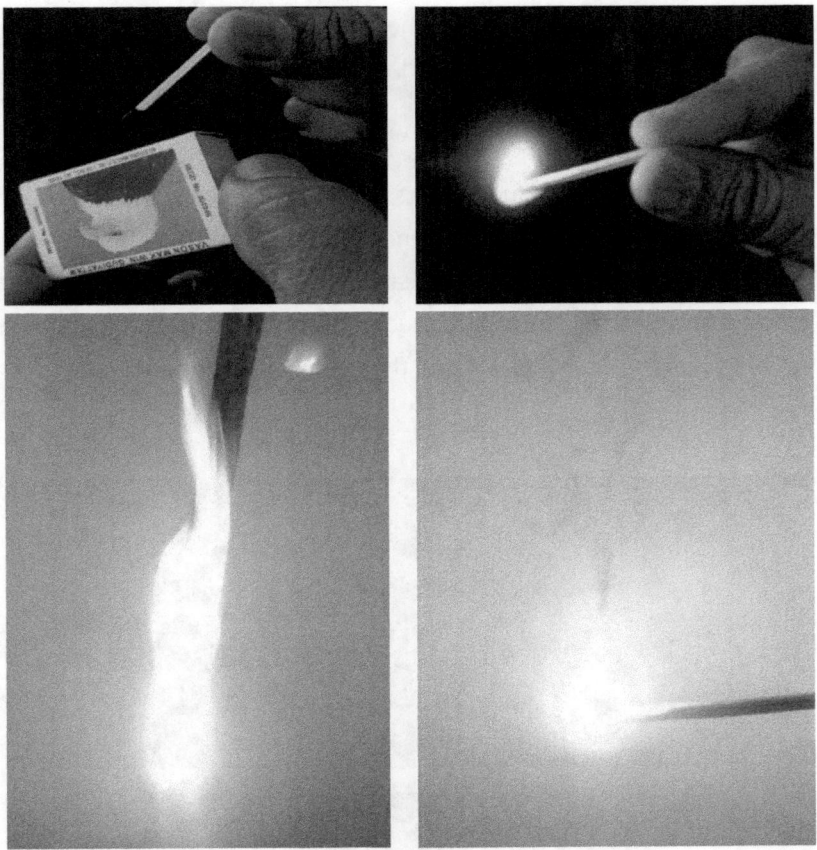

Picture 25 Behaviour of flame - understand through observation

The core of the flame, dark in colour, is where the pyrolysis happens. One can see the pyrolysis gases when the flame is blown out. This is most clearly visible in incense sticks. As long as there is flame at the tip of the stick, there is very little smoke. Once the flame is extinguished, smoke is observed. This experiment is important to explain to the stakeholders the importance of smoke as a combustible gas, otherwise wasted.

As the flame moves towards the hand holding the incense / match stick, the tip gradually turns to ash. After the stick burns out and the flame is extinguished, one can clearly see the ash at the combusted end. The amount of carbon (visible as the black colour formed from wood) reduces gradually from the tip of the match tip (where the flame started) towards the other, un-burnt end.

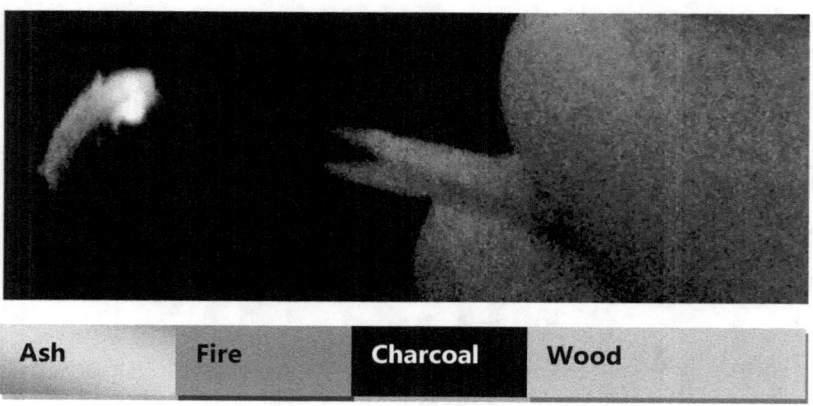

Picture 26 Ash – fire – charcoal – wood: As seen in a matchstick burning

The byproducts of combustion are smoke, moisture, soot, tar, Polycyclic Aromatic Hydrocarbons and Volatile Organic Compounds.

Greater the part of the flame's path that lies in the combustion chamber, more is the chance for the smoke and air to mix and combust completely. Also important in the process is the 'flash point[8]' temperature of the smoke.

> *There could be as many stove designs as the shoes to suite everyone's need*

---

[8] - The flash point: Materials at certain temperature will burst into flames. This temperature is called a material's flash point. Wood's flash point is 572 degrees Fahrenheit (300 C).

Colour of flame

The colour of the flame indicates much about the kind of combustion and whether it is optimal. Reddish flames yield the most soot. The yellowish-red flame is the most common. The bluish-yellowish-reddish flame is optimal.

Picture 27 The sooty reddish flame from pinewood burning

Observing the flames is an experience by itself. One could spend hours just watching them, because the colours are energetic and the flames are dynamic, changing their form continuously. The warmth from the flames is also enjoyable especially during cold weather.

Behaviour of flame

There are different types of flames: rushing flames (as in rocket stoves and chimney stoves), raging flames, forced flames, dancing flames (say as in Magh 3G stove), climbing flames, still or silent flames (wick flames), etc.

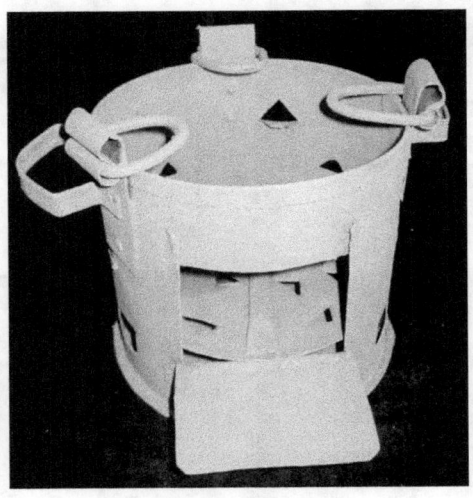

Picture 28 Magh 3G stove with options for using
all types of biomass and charcoal as fuel

Picture 29 Dancing flame

Picture 30 Climbing flame

Picture 31 Raging flame

Picture 32 Rushing flame

Picture 33 Rocket flame

Picture 34 Radial flame

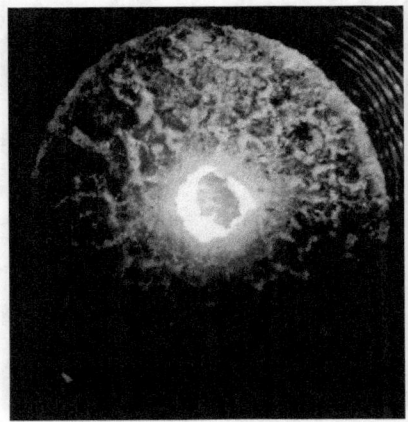

Picture 35 Radiation and flame from briquette

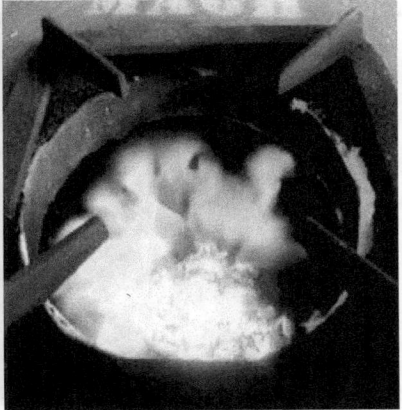

Picture 36 Gasifier flame

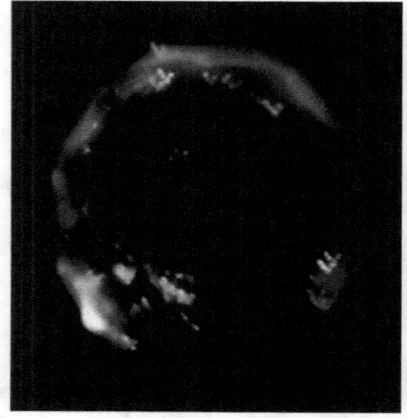

Picture 37 Silent flame – briquette burning with less and constant air supply

Picture 38 Twister flame

Picture 39 Forced rocket flame

Picture 40 Secondary and Tertiary air

Rushing flames are common where there is a chimney effect, or a chimney. In a rushing flame stove, although the flames appear dynamic and impressive, the conduction of the heat to the pot is low. In a dancing flame stove, air mixes thoroughly with the combustible gases, and the flame takes its time to combust gases without rushing below the pot. Therefore, the heat conducted to the pot is high. In forced air stoves, the flames are forceful; one can even listen to the sound of the flames quite clearly. Forced air flames hit the pot and are deflected immediately. The rate of absorption of the heat from these flames is lower than dancing or still flames. Sometimes, due to excess air in the forced air stoves, there could be a cooling effect on the pot. Silent or still flames are observed in wick stoves or lamps. Under calm conditions, the flames in a biomass stove could be still or silent. The silent flame is a result of a continuous release of combustible gases and constant mixing of air. On the other hand, a raging flame is the most uncontrolled. It is formed when too much fuel is added and the air mixes uncontrollably.

Thermal aspects

In a stove, there are three ways in which heat is transferred to the pot and its contents.

> ☐ On thermal aspects of flame – Conduction, Convection and Radiation.

Conduction

When the flames touch the pot directly, the mode of heat transfer is called conduction. The rate of transfer of heat through conduction is the highest. People commonly believe that food can be easily cooked only through conduction. In many traditional stove designs, the height of the stove is kept very low to enable efficient conduction.

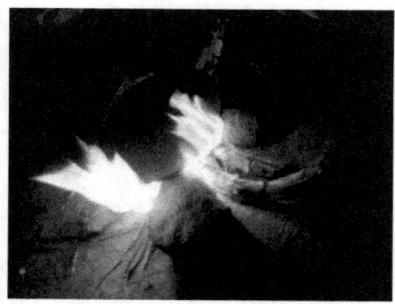

Picture 41 Traditional three stone stoves are of low height

People are not easily convinced that, in good stoves with efficient designs, heat transfer can happen through convection and radiation as well. During conduction, the flames embrace the pot sideways at times, and could reach even higher than the pot (as often seen in low-height three-stone stoves). When that happens, very little black soot is left on the pot, as the excessive heat burns most of it into ash. Ash is white in colour; it does not stick to the utensils and therefore the pot surfaces appear relatively cleaner.

In the traditional low-height stoves, black soot appears mostly on the kitchen walls and less on the pots. In the taller, efficient stoves, little soot is formed on the walls, but some soot is found on the pots on the outside. On stoves with chimneys, the flames do not rise and embrace the pot as high as they do on traditional stoves. The soot particles, along with the wood vinegar / tar, tend to stick to the pot. However, overall, the amount of black soot formed on kitchen walls due to efficient stoves is very less. On 'cool pots', the chilling effect also contributes to the deposition of soot. If the pots are smooth, less soot is deposited on them.

Fire wants to keep away from coming in contact with surfaces, and move forward. Therefore, it is important to keep the walls of the combustion chamber straight and smooth to minimize heat loss due to conduction. Many people focus more on what the stove looks like from the outside, rather than the inside. When constructing a stove, especially using materials like bricks and clay, one should make sure that the walls in the combustion area are smooth on the inside.

Convection

Convection currents within a stove always move from bottom to top. They are formed due to pressure-difference: high pressure at the bottom of the stove and very low pressure at the top. This difference in pressure is, in turn, due to the difference in temperatures: high temperature right below the top creates an area of low pressure and colder, heavier air at the bottom of the stove creates high pressure. The convection currents support the mixing of the air with combustible gases inside the stove and sustain the fire.

Sometimes, the cold air in the stove has a dampening effect on the fire. Part of the energy is also consumed for preheating the air before it can join the combustion / pyrolysis zone. Preheated air is easily combustible. All the air streams (Primary air, Side air, Secondary air and Tertiary air) support the formation of convection currents. The intensity of the currents is reduced by the drop in pressure due to air flowing in from the side and secondary sources. In Magh 3G stoves with primary air, side air and secondary air sources, the flame is dancing rather than rushing.

If the convection currents are stronger after the initial burning of the fuel, and when the stove is hot, the size of the air holes is not a major concern. In that case, more air is easily sucked into the flame. When running a stove on high power, the excess intake of air forms into a strong convection current. This reduces the efficiency of the stove. In windy conditions (that add to the draft), and in forced draft stoves, the convection currents are stronger. It is due to convection currents that a stove continues to burn when started.

Many traditional stoves have to be warmed up. This is done usually by operating the stove at high power for some time. Once a stove is extinguished, it has to be blown into to rekindle the flame. The blowing is also meant to initiate the convection current. The primary air, preheated by the charcoal embers at the bottom of the grate, is crucial for continuous combustion in the stove.

Radiation

Radiation is another mode of heat transfer that many stove users do not understand easily. One of the best examples to explain it is how, without any media, heat is transferred between the sun and the earth. By placing hands around the stove, one can feel the heat from the stove; that is radiation. In less insulated and metal stoves, the radiation is more easily perceived.

Heat-loss due to radiation occurs in all directions. Radiation also transfers heat to the bodies of those cooking on the stove. In places with colder climates, it is radiation through which interiors of homes are heated.

Cooking food is a culture and the recent adoption of modern cooking methods is not a substitute for traditional biomass stoves

No Smoking

"Smoke is smoke" and too much exposure to smoke is always bad. In many countries, governments have banned smoking in public places. When millions of people are being killed due to indoor air pollution globally, why is smoke not banned inside every house? Sometimes, exposure to smoke released by cooking stoves every day is more harmful than smoking 100 cigarettes. There is, of course, a difference: cooking is a compulsion and exposure to smoke from inefficient stoves is a hazard, whereas smoking cigarettes is a habit and voluntary.

Picture 42 Similarly the campaign for good stoves is needed

Overall, kitchens are not the highest priority for most people. Kitchens are smaller than other rooms in a house. The money spent on constructing a kitchen by

☐   On Smoke, Soot and the emissions from cook stoves.

poor people is always minimal, and usually less than the amount they spend on toilets. Even in government housing schemes, kitchens are not given priority.

Of late, many youth have inquired about biomass stoves. They have wanted to give one to their mothers / grandmothers, who are still cooking on smoky, inefficient ones. Many older women in rural areas are afraid of Liquefied Petroleum Gas (LPG) stoves, which they think are dangerous to operate.

Picture 43 Children usually close to their mother while cooking are exposed to smoke

Cooking is a compulsion and exposure to smoke from stoves is an inevitable hazard, whereas smoking tobacco is a voluntary habit and for pleasure

Soot and smoke

Pieces of carbon left in the stove after cooking is called charcoal. Soot is made up of very small pieces of carbon flying out from a fire. Smoke includes soot, moisture, combustible gases Polycyclic Aromatic Hydrocarbons (PAHs) and Volatile Organic Compounds (VOCs). Soot is seen on the stove, the utensils used for

Picture 44 Soot accumulated on the walls of kitchen

cooking and on the kitchen walls. It is a highly undesirable substance: black soot in the air contributes towards global warming. Black soot that reaches and is deposited on ice caps reduces their 'albedo', or reflective power. With a lowered albedo, ice absorbs the heat that it could have otherwise reflected, and melts faster.

Picture 45 Soot accumulated on the roof of a kitchen

Picture 46 Circular frames made of clay / metal used for prevention of soot on cooking utensils

People do not take pride in their kitchen. It is the last place in the house shown to a visiting guest

Emissions

The pyrolysis and combustion of wood in biomass stoves produce atmospheric emissions of particulate matter: carbon monoxide, nitrogen oxides, organic compounds, mineral residues and, to a lesser extent, sulphur oxides. The quantity and type of emission is highly variable, depending on a number of factors including the stage of the combustion cycle. During the initial stages, after a new wood charge is introduced, emissions (primarily VOCs) increase dramatically following the initial period of rapid burning. There is also a charcoal stage in the cycle, characterized

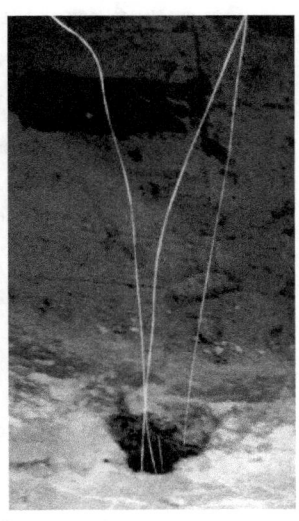

Picture 47 Shooting soot particles from a stove

by slower burning and lowered emissions, but invisible and undetectable carbon monoxide emissions are then high. Emission rates during this stage are cyclical, characterized by relatively long periods of low emissions and shorter episodes of emission spikes[xv].

Air for combustion

Depending upon the point of entry, air entering a stove can be classified as Primary Air, Side Air, Secondary Air and Tertiary air,

> ▢ On sources of air for efficient combustion – Primary air; Side air; Secondary air; Tertiary air. Types of burners, concentrators and air controls. Grate and its functions.

Primary air

This is the air entering the stove mainly from the bottom of the grate / underneath the biomass fuel. The excess flow of primary air will lead to uncontrolled combustion. Primary air is always required in low amounts. If it is preheated before reaching the fuel, the combustion is efficient and the stove performs better. After the preliminary combustion, the embers at the bottom of the grate serve to

Picture 48 Primary air enters underneath a grate

preheat the primary air. In TLUD stoves limited amount of primary air entry leads to the release of woodgas.

Side air

In a majority of the traditional stoves, this is the only source of air. Side air gushes in through the fuel feed opening. In 'three-stone' stoves, air gushes into the combustion zone from all three sides. Too much air streaming into the stove (for example when it is situated outdoors, in the open) has a dampening effect on the flames. Controlling the gushing side air is very important. In Magh 3G stoves, a shutter controls the entry of side air. If the mouth of the fuel feed is completely closed due to overfeeding of fuel wood (as could happen in traditional stoves), the flame could be extinguished and release excess smoke. Small holes on the sides, as provided in Magh 3G, also keep the stove from extinguishing. In good stoves, the fuel feed opening should be made smaller, as there is already scope for primary air to enter through the grate.

It is easy to design a stove for a specific fuel

Secondary air

Use of secondary air is an important feature of TLUD stoves. This feature can also be incorporated in any stove (with fewer holes than TLUDs); it will help the complete combustion of smoke. In stoves without chimneys, this is an especially useful feature. Preheated secondary air helps in clean combustion of the gases. The secondary air can be preheated by passing the air through the hot parts of the stove before it enters the combustion zone.

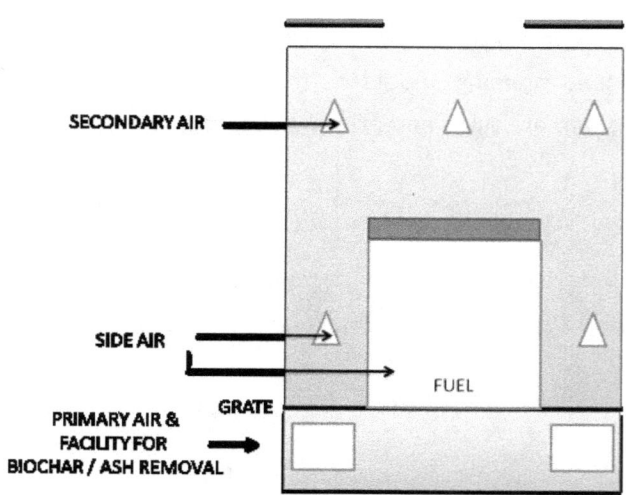

Picture 49 Source of secondary air, side air and primary air in Magh 3G stove

Tertiary air

Tertiary air helps in complete combustion of the gases and drastically reduces emissions.   Preheated tertiary air is more beneficial than cold tertiary air. This is a rare feature, not found in a majority of stoves.

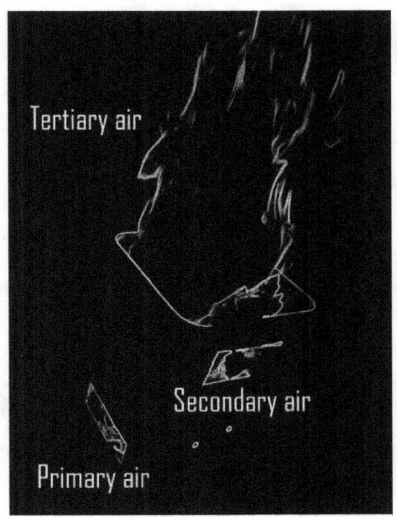

Picture 50 Tertiary air enters after
secondary air entry into combustion zone

Burners

Burners are the components in stoves that ensure complete combustion of fuel. In stoves with burners, there are four types of air entering the pyrolysis / combustion zones:

T - Tertiary air - >

S - Secondary air ->

                                                     <-Si - Side air

P - Primary air - >

Burners, pot rests, grates, ash containers, handles and safety screens etc., are important components of stove design.

The burners shown below can work with forced air or natural draft air.

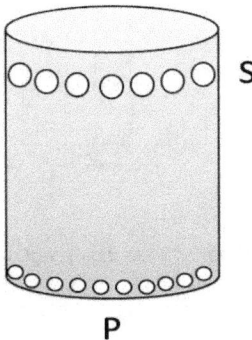

The primary air holes from the sides are suitable for small size TLUDs.

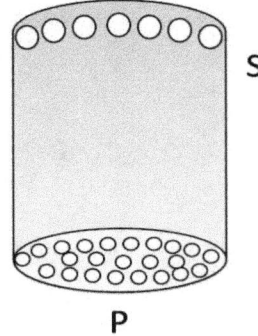

The primary air holes at the bottom are suitable for large size TLUDs.

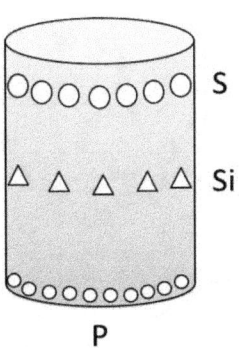

Side air is a feature for largesize and tall TLUD stoves, in which the biomass is fed from the side.

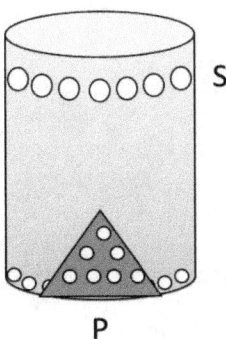

The central conical primary air facility is suitable for large size TLUDs.

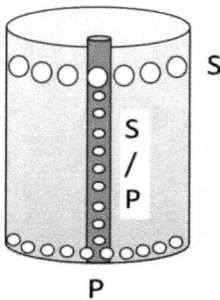

The air from the central pipe also a source of primary and secondary air.

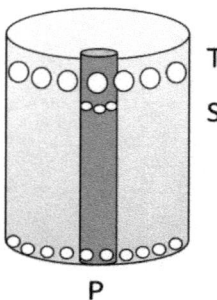

The pattern of holes incorporating flow of three types of air - Primary, Secondary and Tertiary air for complete combustion.

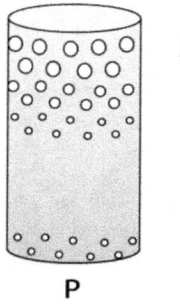

Gradient secondary air holes are suitable for Tall TLUDs, which provide the primary air initially until the biomass collapses after combustion.

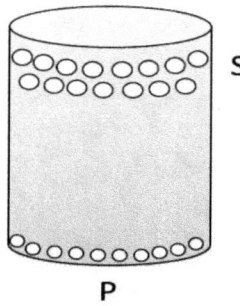

Double secondary air holes are suitable for big-size TLUDs.

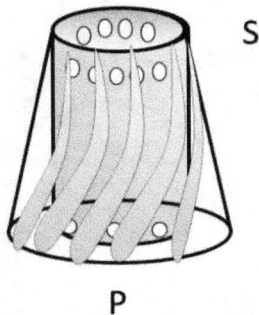

Option of fins in the burner for preheating air - through surface contact, increased surface area and increasing the speed of air entry.

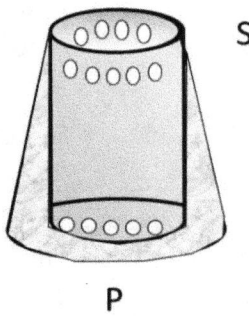

Steel wool around metal combustion chambers for preheating the air.

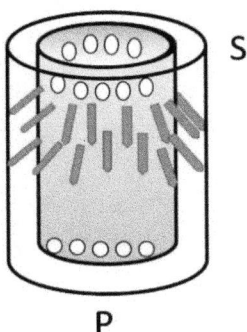

Fins for preheating secondary air.

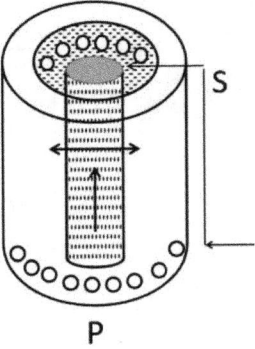

The very small holes used for the central pipe for primary and secondary air.

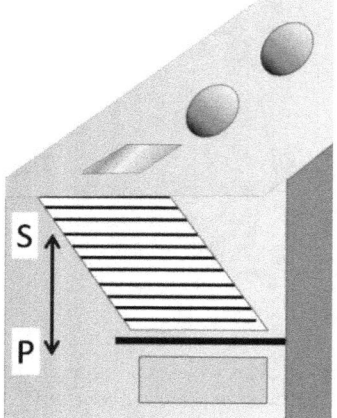 

Primary air, secondary air and side air flow into combustion zone of the stove. This component also acts as a slide – loose biomass fuel such as wood shavings, groundnut shells, etc., can be fed in.

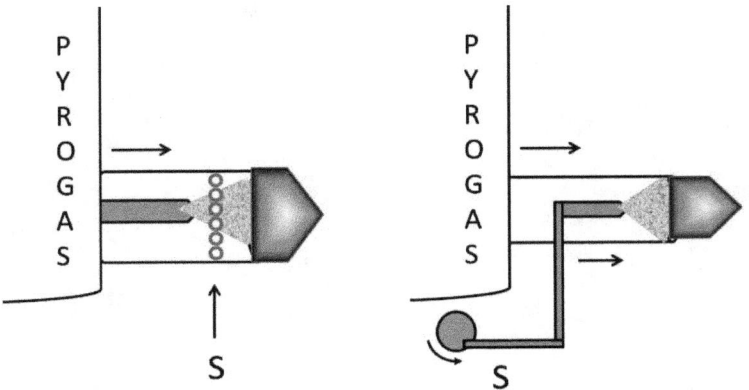

The pyrogas from a gasifier is forced out and secondary air enters naturally mixes and combusts.

The secondary air is forced. It sucks the pyrogas from the gasifier and mixes well before burning.

The burner need not be upwards. By rotating the burner, one has options to incorporate different types of gasifiers[9].

---

[9] Gasifier is the technology for Gasification (for example Top Lit Updraft Gasifier stove / woodgas stove). Gasification is a process that converts organic or fossil-based carbonaceous materials into carbon monoxide, hydrogen and carbon dioxide. This is achieved by reacting the material at high temperatures (>700° C), without combustion, with a controlled amount of oxygen and / or steam. The resulting gas mixture is called syngas (from synthesis gas or synthetic gas) or producer gas and is itself a fuel.

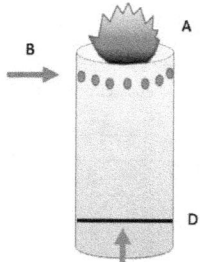

Simple TLUD batch load gasifier – loaded from top.

Side load gasifier.

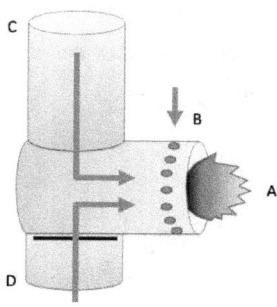

Top load gasifier

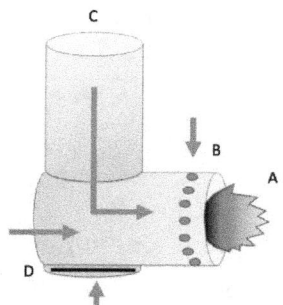

Top load gasifier

A – Fire, B- Secondary Air, C-fuel feed, and D- Primary Air and grate for ash removal (Note: Arrow directions indicate air draft also) .

Concentrators

Concentrators are simple obstructors that help mixing of air and gases and ensure complete combustion. They are of various types. Simple concentrators constrict the flow of gases into a small hole. Some concentrators achieve this by using of a number of small holes.

Picture 51 A simple tin stove with a concentrator

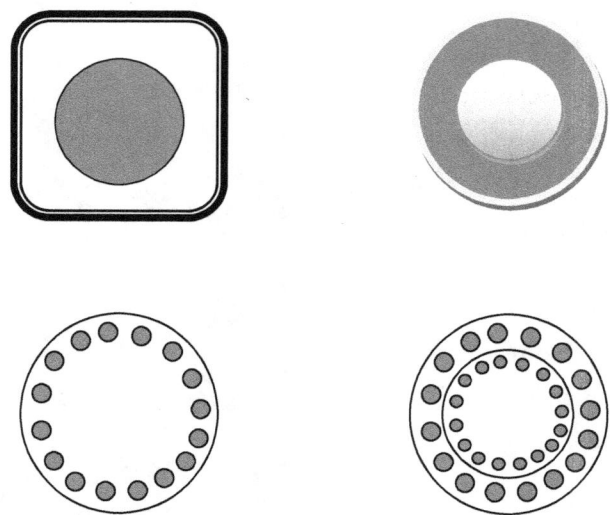

Picture 52 Top view of concentrators - red colour is the fire

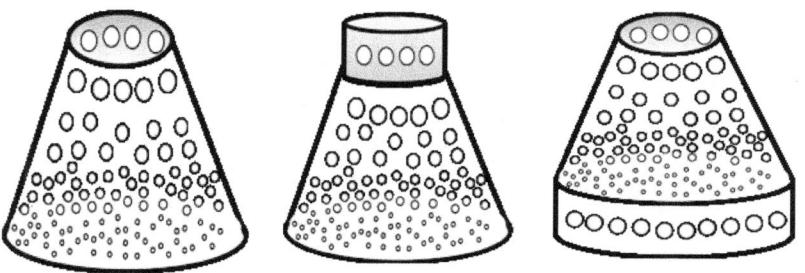

Picture 53 Conical concentrators – Side view

Air controls

With scope for regulated entry of all the kinds of airs - primary, side, secondary and tertiary, a stove can be used easily even outdoors. By controlling the entry of air into the combustion zone, any stove can be used to its maximum efficiency. With enough experience and understanding, one can cook even on a three-stone stove with great efficiency.

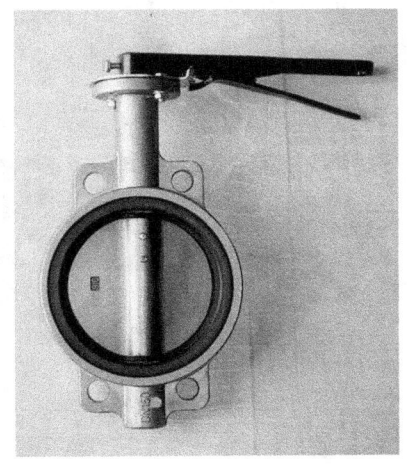

Picture 54 Butterfly valve for controlling airflow

For efficient combustion, the use of primary air should be minimal as compared to side and secondary air. When a stove runs at high power, excess draft is created and flames are intensified. This reduces its efficiency. In stoves with chimneys, the excess draft created due to the chimneys reduces their efficiency. By having air controls within the chimney, the efficiency of the stove can be improved. Butterfly valves can be inserted in the chimneys for this purpose.

Institutional stoves should have thermometers installed in them, because it is otherwise difficult to observe the temperatures within the stove. Two thermometers – one at the bottom of the pot and another at the exit point in the chimney — are preferable. A dampener (for example Butterfly valve) can be inserted into the chimney for controlling the heat lost through it.

By careful observation and monitoring measurements, one can tame the flames as desired

Grate and stove bottom

In domestic stoves, the gap between grate and bottom of a stove could be around 1.5 to 2 inches. Seldom do people remove ash while cooking. Ash being a good insulator protects the bottom of the stoves from overheating. The embers falling beneath the grate help to preheat the primary air entering the stove. However, the ash has to be removed from time to time to make sure that the movement of the primary air is not blocked.

In institutional stoves, this gap can be sized as per the specific needs of the establishment. In a stove used for cooking for 200 people, the gap should be around 6 inches. In all cases, a mechanism to control the primary airflow helps efficient combustion. The air control could be in the form of a sliding shutter, a door with holes, a sliding door, a butterfly valve, etc.

Pots and efficiency

The efficiency of a stove is the amount of heat (energy) that needs to be transferred through it for the ultimate purpose, which is cooking the contents in a utensil. It should be noted that heating the utensil is not the ultimate purpose. The same stove could

> ☐   On pots and their efficiencies – heat transfer, special features to pots for heat transfer, size of the pots, pot rest, air gap and skirts for pots.

exhibit different efficiencies under different conditions. Therefore, one needs to consider the average efficiency as its defining feature. One of the main factors affecting the efficiency of a stove is its mass.

There are more toilets displayed in museums than stoves

Picture 55 Stoves designed by Dr N Sai Bhaskar Reddy at "Stoves Museum", GEO Research Center, Peddamaduru Village, Devaruppala Mandal, Warangal District, Andhra Pradesh, India

Heat transfer

Pots made of material such as aluminum or shining steel reflect part of the heat transferred to them. Some metals – such as steel, aluminum and copper - have a higher capacity for heat absorption and transfer as compared to earthen pots traditionally used by the poor. Sometimes, people apply soil to the pots in order to increase their heat absorption capacity. Another important advantage of this practice is that it minimizes the amount of soot accumulating on the utensil, making it easier to clean. Applying soil to utensils is a common practice in many parts of India.

Metal pots that transfer a large proportion of the heat received to their contents, do not do so uniformly. When stoves run

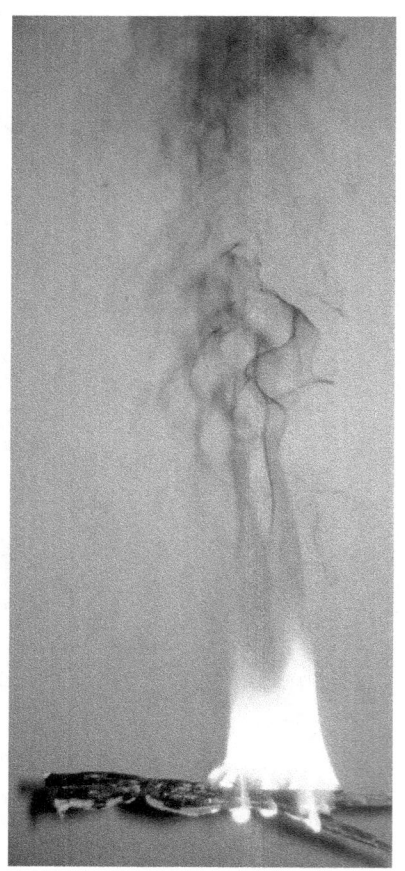

Picture 56 Soot released from burning of pine sticks

at high power, the food is not cooked well. The flavour, taste and aroma of the food also depends on the temperature at which it is cooked. The temperature range 300 – 600 degrees centigrade is ideal.

Cooking is also a means for releasing the nutrients of the food, in order to maximize its value to the consumer. At higher temperatures, the aromatic ingredients of the food are also lost as volatiles.

Cooking food is a culture and tradition developed through the use of biomass stoves over centuries. Modern means of cooking do not achieve the same taste, flavour and nutrient value as cooking on biomass stoves. Some of the common ways of high temperature cooking used today are: LPG stoves, microwave ovens, electric stoves, induction stoves, etc. High pressure and temperature cooking is done by using pressure cookers.

During high power cooking, metal pots made of aluminum and steel transfer excessive heat to the food, and not uniformly so. The food cooked in earthen pots is more tasty and nutritious, due to the slow and low transfer of heat to the food that they facilitate. However, this is not the preferred mode of cooking now-a-days. New cooking methods have to adapt to modern sources of fire / heat.

Foils for pot

To certain utensils, which are fixed to the stoves, thin metallic foils can be fixed to increase their efficiency. Many sugarcane farmers produce *jaggery* as a household-level business. The sugarcane juice is boiled in large shallow pans fixed to large-sized biomass stoves. The heat transfer to the pans increases due to thin metal foils fixed to them.

Of the cost of a house, a good stove costs one to two per cent and a toilet five to ten percent

Pot sizes

Utensils used for cooking come in different shapes and sizes: cylindrical, tapering from the middle of the pot, semi-spherical, bowl-shaped, etc. Pots of the same volume but different sizes (height versus diameter) and shapes have different efficiencies. A tall pot is less efficient than a

Picture 57 While cooking rice on a large scale, embers are placed over the lid at the end

wide-bottom pot of the same volume. Similarly, pots with spherical bottoms are embraced by the flames more easily. More air, then, can be drawn into the flames as they climb up along the sides of the pot. Thus, heat is transferred not only from the bottom, but also from the sides. As the pot-rests are fixed, when round bottom pots are used, there is an excess air gap and the efficiency of the stove comes down. For making curries, pots with hemispherical bottoms and tapering tops are most suitable. Cooking curries require frequent stirring. Therefore, due to the tapering tops, the loss of heat from such pots is low. For cooking rice, flat-bottom cylindrical pots are used. In cooking rice, water is the main medium. Therefore, cylindrical, flat-bottom, low-height pots are able to distribute heat uniformly. For large scale cooking, flat or slightly convex bottom pots with tapering mouths are preferred. In a big utensil, the rice at the top often stays uncooked, so people often place charcoal embers on the lid after cooking.

One observes that hemispherical / curved / almost flat pans are used for frying, roasting, baking etc. They are open and not very efficient. For boiling milk, tall utensils are preferred. They are less efficient and lose some heat due to radiation, but their shape keeps the milk from boiling over.

Promoting less efficient cooking stoves knowingly is a crime

Pot rest

Pot-rest is one of the most important components in a stove. The durability of the pot rest is very critical, because the pot has to bear all the weight of the utensil with food. Parts of the pot-rest are subject to the highest temperatures in a stove during cooking. Where large size pots are used, they are inevitably dragged

Picture 58 Metal Pot rest

over the pot-rest. To withstand all that, it has to be one of the strongest and most durable parts in the stove.

In a three-stone stove, the portions of the three stones touching the pot act as the pot-rest. Domestic three-stone stoves are relatively small in size, and mostly made of stones (or sometimes bricks). In clay and mud stoves, small pebbles / stones are used as the pot-rest. Often, women maintain and mend stoves using plaster made out of clay and cow-dung. They do not

Picture 59 Pot covered outside with a layer of clay, for easily cleaning the soot after use and to increase efficient heat transfer

forget to plaster over the pot-rests as well. Pot-rests made of forged iron, wrought iron or iron are also used widely; they are more durable.

In some stoves, a dummy circular frame rests above the pot-rests. This frame prevents soot from forming on the sides of the utensil, besides acting as a pot rest by itself.

Picture 60 In large scale cooking the pots are sometimes dragged – often leading to damage of pot rests

Pot Rest Designs

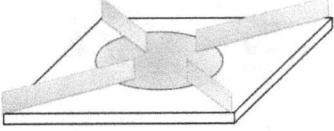

Simple pot rest

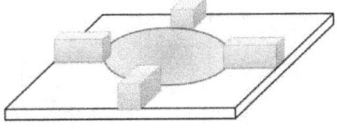

Simple pot rest

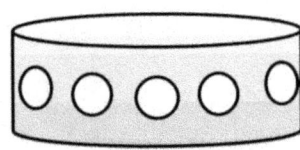

Pot rest with circular holes

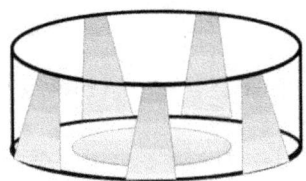

Pot rest with gaps

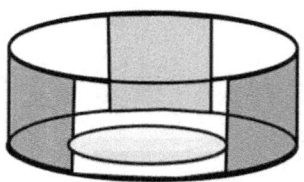

Pot rest with gaps

The radial pot rest sitting on three stones, increases stability of the pot

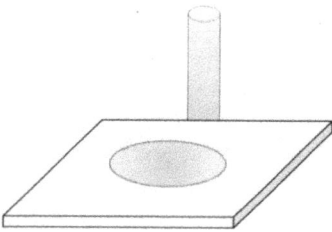

Pot rest with a chimney

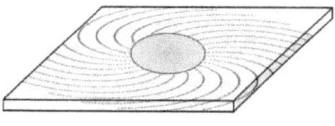

Radial movement of heat and fire, to increases the surface contact with the pot and also reduces emissions

Air gap

In all non-chimney stoves, an air gap is required between the stove and the utensil. This allows a draft, without which the flame will not move upward and make contact with the pot. Too much or too little air gap reduces the efficiency of the stove. In domestic stoves, the air gap is about one inch. In a charcoal stove, the gap between the fuel and utensil should be even less, since there are no aggressive flames and cooking happens mainly through radiation and convection.

Flames need to accelerate before hitting the pot for efficient transfer of heat to the pot. Too much cold air entering the combustion zone sometimes dampens the flame and leads to smoke. The ideal air gap also varies with the shape and size of the pots.

In general, almost all the pots used for cooking are cylindrical with circular bases. In keeping with this shape, the combustion chambers are usually cylindrical as well.

Skirts for pots

Although skirts improve the efficiency of stoves, they are rarely used in households. With well-designed skirts, the air gap can be maintained perfectly, and the heat can be forced onto the sides of the utensil. When the cooking is simple-- say with few utensils-- it is convenient to have a skirt for each pot. When pot shapes and sizes are diverse, flexible pot skirts can be used.

In practice, pot skirts are used mostly in institutional kitchens at big establishments: such as schools, restaurants, etc. In such kitchens, the pots are sometimes fixed to the stoves and the perfect air gap is maintained.

A pot is more stable on three stones rather than on four stones, therefore there are "three-stone stoves" being used widely

Chimney stoves

Chimney stoves have greatly reduced exposure to indoor air pollution. Any stove with a chimney reduces emissions considerably. The cooks' exposure to heat is also greatly reduced, because in chimney stoves there is no air gap between the pot and stove. Sometimes, chimneys without dampeners create excess draft, leading to excessive combustion of fuel. The placement of a chimney does not make the smoke disappear completely. Sometimes, the smoke coming out of the other end of the chimney is an area of concern.

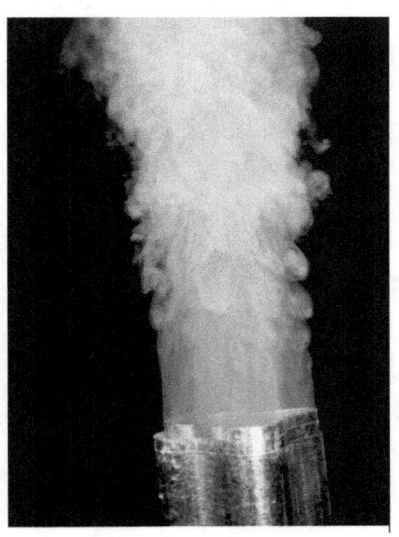

Picture 61 Smoke including moisture from an institutional stove

Chimney stoves are not necessarily more efficient. In fact, most often, they are less efficient. Their rushing flame does not conduct heat to the pot very well. Moreover, chimney stoves are

Picture 62 Butterfly valve used in a chimney

expensive. In the case of domestic stoves, the chimney sometimes costs as much as the stove. Besides, their upkeep is difficult. It takes a brush with a very long handle to clean a chimney. In villages, people use leafy branches from trees. If not cleaned frequently, the chimney is blocked with soot, which reduces the stove's draft. Many a times, villagers fail to maintain their chimneys, which are then removed and used for other purposes (as drainage pipes, for example). At times, the heat lost in chimney stoves is as much as the amount of heat required for cooking!

Concrete chimneys are less durable. They have to withstand very high temperatures. Any stove design with a chimney can be sold as a good stove with low indoor air pollution, but need not be efficient or pollution free. In cold countries, the excessive heat channeled through the chimney is used for heating rooms / water and warming the beds, etc.

Special features in stoves

There have been efforts to design stoves in which steam or atomized water is introduced into the burning fuel, to split hydrogen from water to help combustion. Very few stoves have this feature and the increase in efficiency is very low. The amount of energy required for splitting water (H_2O) into hydrogen and oxygen is very high. It is only at 2200 °C that about three percent of all H_2O molecules are

Picture 63 A metal tube coil for steam generation used in a stove

dissociated into various combinations of hydrogen and oxygen atoms - mostly H, H_2, O, O_2, and OH. Nevertheless, the idea of incorporating multi-fuel mixes is in its infancy and we will most probably see more such designs in the future.

Water will be an important source of fuel in times to come. When steam is added in a blue flame, orange / pinkish flames are visible. Some people have observed increased efficiency in a stove upon introduction of steam. The increase in efficiency could be most likely due to the decrease in the rate at which combustible gases are being formed in the stove. Most biomass stoves generate heat between 300 - 700 degrees centigrade under the pot, and the amount of hydrogen gas generated from the steam would be very little.

Thermoelectric generators (also called thermogenerators) are devices that convert heat (temperature differences) directly into electrical energy. This phenomenon is called the "Seebeck effect" (or the "thermoelectric effect"). Their efficiency is typically around 5 – 10 %. The University of Nottingham has designed a biomass stove with thermoelectric power generation capability, called SCORE (Stove for Cooking, Refrigeration and Electricity).

Charcoal from stoves

Many stoves achieve higher efficiencies when the charcoal continues to remain in the stoves (including TLUD stoves). People are reluctant to remove charcoal left behind after previous cooking, and continue to cook. In such cases, the stove functions as charcoal stove. The charcoal yield varies across types of stoves. For example, in Magh 3G stoves it is about 15 % by weight, in TLUD stoves the yield is about 25 %.

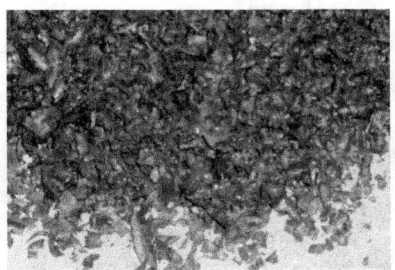

Picture 64 Charcoal yield from Magh-1
gasifier stove

Sometimes a scientist is successful in the lab, but fails when he goes out to the community

Picture 65 A girl cooking on an efficient stove in the open

Facilitation of domestic stoves

Facilitating good stoves requires a number of steps:

Reconnaissance: This involves a study of existing stoves, biomass and other fuels being used in a geographical area. Without reconnaissance and a preliminary understanding, stoves should not be recommended directly.

 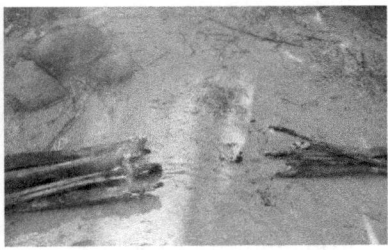

Picture 66 Demonstration and comparison of traditional and efficient stoves

Picture 67 Seeing is believing – comparing the wood saved using traditional and efficient stove

Awareness and sensitization: Intensive awareness and sensitization should be created at the community level regarding biomass conservation, and about the need to adopt efficient stoves. The sensitization creates a sustainable motivation among the community. Awareness can be created through various means: Wall writings, posters, skits, plays, workshops, demonstrations, meetings, presentations, films, videos, press, etc.

Design: The stove design should be simple, preferably made using locally available raw material. Testing stoves and subsequently improving their design is an integral part of the design process.

Picture 68 Women with least skills can be easily trained on certain stove designs

Demonstration: Explaining the stove's design and performance with the community through village-level demonstrations / workshops is great for eliciting community participation and awareness.

Piloting: One should select leaders from the community to help with piloting the stove, creating awareness, and as a source of an overview of the community feedback based on which the design could be improved.

Scaling Up: Training youth, women, artisans, masons and so on, in the construction and maintenance of good stoves would help scale up their demand and supply. By creating linkages and converging with various agencies, funds from governments, donors and other organizations can be accessed. This would also help disseminating the stove over a large area and expanding demand.

Monitoring: Monitoring the performance of the stoves should be a continuous effort. Therefore, maintenance and related services should be offered by trained local people.

Achievements / Results: Documentation of the number of good stoves being adopted by communities, by location.

Recognition: Recognizing all the stakeholders involved directly or indirectly in the promotion of good stoves – members of the target community, facilitators, donors, other support organizations, scientists and institutional partners involved in the process.

Escalating prices of a stove for the sake of business or schemes affects the purchasing power of the poor

Institutional Stoves

Institutional stoves, also called community stoves, are designed for large scale cooking. Domestic cooking stoves, on the other hand, typically meet the cooking needs of a family.

Picture 69 A makeshift three stone stove used in a school

A majority of the institutional stoves currently in use are three-stone stoves, which are highly inefficient. These stoves require less space and can be moved easily to any place as per convenience.

In institutional stoves wood is used as fuel more often than in domestic stoves, in which households often use crop residue / twigs branches / etc., very often. This type of use has potential affect on the environment.

☐ On Institutional stoves – preparations and installation of institutional stoves; stoves efficiency; and adaptation factors.

Even though efficient stoves for institutional kitchens appear more expensive initially, the savings made on wood over a period make them less expensive overall.

Picture 70 Hazardous way of lifting and bringing down the cooked food from stove

Picture 71 An efficient institutional stove

Picture 72 Testing an institutional stove in a school

Preparations for installation

Some of the important activities that need to be carried out before installing efficient stoves are:

- Preliminary study of the existing stoves;

- Sharing observations with the decision makers, and *Picture 73 Institutional stove* emphasizing upon the importance of adopting efficient stoves;
- Creating awareness among stakeholders;
- Collecting information regarding existing facilities for cooking. This information is critical. For example, sometimes the space available for installing stoves is not feasible (less area, bad ventilation, open air conditions, prone to the vagaries of weather, water seepage, etc.);
- Planning the number of stoves required;
- Deciding on the size of the stove as per the cooking needs and utensils being used;
- Ordering the material required for the stove and preparing the material for the stove;
- Discussing with the main stakeholders and deciding upon a convenient day for the installation. It should not hinder daily cooking. If needed, a makeshift stove should be installed, for use while the efficient stove is being installed. Ideally, the installation

of the stove should be completed within a day. Again, the idea is to affect the daily business of the institution as little as possible;

- The place chosen for installation should be marked a day in advance;
- Installation parts should be prepared well in advance (prefabricated). They should be sent with other necessary material to the place of installation at least a day in advance
- The team responsible for installation should be intimated well in advance;

Installation of stoves

On the day of installation of an institutional stove, one should arrive with all the required tools and equipment: Crowbars, spade, plumber's thread, plumbing bob, level meter, steel scale, tape, hammer, etc. It helps to prepare a checklist in advance.

Picture 74 Installation of institutional stove

The mason, labourer and other support persons should be intimated one day in advance about the time and the site of installation.

The team lead should have with him a plan of the place, stove design, a camera, etc. The role of team lead is very critical; all the decisions have to be made instantaneously. In addition, any challenges emerging during the installation need to be resolved on the spot.

Picture 75 Using an utensil with water (level) for installation of a stove parallel to the ground.

After installation of a few stoves the supervisory role of team-lead reduces, because the masons and other support staff can thereafter carry out most of the work themselves. The installation team should stay unchanged throughout the installation process.

The installed stove should be tested extensively, as soon as possible (roughly within three days). If any problems are observed, they should be rectified during that period.

The chief kitchen staff of the institute / establishment should be involved right from the planning stage, through the choice of location, right down to the installation and testing of the stove. If needed, they should be trained in operating the stove. They would be the main user and should be absolutely satisfied. If not, the stove could be a failure despite being highly efficient.

Sharing is Freedom and Integrity

Stove efficiency

The efficiency of the stoves depends mainly upon the following aspects:

Utensil: If the cooking utensil used is large in size, some heat will be lost through radiation.

Stoves: Efficiency-reduction is due to radiation, convection and body mass of the stove.

Fuel wood: Moisture content and type of the biomass fuel used.

The skills and proficiency of the cook.

Heat losses occur mainly through radiation from the stove, convection around the stove and convection through the chimney (in cases where chimneys are used). Besides, a large amount of heat is used up in heating up the stove body and the fuel (especially if the fuel is moist), before the heat can be utilized for cooking. Heat is also lost unproductively when ash / charcoal that is still hot is removed from the stove.

Stoves' efficiency can be improved by:

Insulating the stove wall.

Good stove design – in terms of enabling the required amount of airflow through it.

By using pre-dried fuel of the suitable size (fuel pieces which are too large being particularly inefficient), and feeding it at an appropriate rate.

Ensuring that the stove is not much larger than needed.

Training the cook in using the stove efficiently facilitating good stoves for institutions.

Adoption factors

By adopting efficient stoves, institutions can save about 30 to 40 % on their wood consumption. These stoves can be designed using common bricks, refractory bricks, insulation bricks, refractory clay, ash, cement, sand, metal, steel frame and iron rods grate, etc.

Sometimes, facilitating the adoption of a good stove by a new institutional kitchen is easier than for the existing institutions. Here are some of the factors that affect the process:

Existing Space: Efficient 'good stoves' often consumes more space than the traditional, inefficient stoves. Apart from space for the stove itself, there should be enough space for the cooks to move around.

Cooking space: A space is needed for keeping – cooking ingredients within the kitchen. Such material should be easily reachable and kept close to the stove. Space for utensils and cutlery such as spoons and so on, is also important.

Firewood in the kitchen: There is a need for a space to keep a certain amount of firewood needed during cooking. Those who are cooking would rather not go out to get it again and again. They might also prefer to stock up all the fuel wood needed for a long period.

Cooking spaces: Cooking is done in different types of spaces: a) under trees (completely in the open) b) in semi-ventilated conditions (small, 3 to 5 feet walls enclose the cooking area which often has a thatched/ tiled / tin / asbestos roof) c) in a closed environment (rooms with hood, chimney, window/s, ventilator/s, doors, exhaust etc.)

A place for pre-cooking activities: such as chopping vegetables, cleaning rice and pulses, etc. This place is very important and should be close to the cooking area. Due to the intense heat that cooking stoves generate, such pre-cooking activities are carried out far away from stoves. Much fuel and time will be saved if this place is close to the kitchen.

Food Storage: For supplies that are brought in on monthly / fortnightly / weekly basis (as per convenience). The storage room should also be located close to the kitchen.

Location of the water-source: Water is an important requirement for cooking. If piped water supply is available close to the kitchen, a hosepipe can be used to bring water to the cooking area. Where running water is not available, a drum / big storage vessel is kept within or very close to the kitchen. Such containers take up a lot of space and make it inconvenient for the kitchen users to move about.

Place to wash utensils: All utensils, including plates and cutlery need to be washed before and after cooking. This place is often found close to the kitchen. If not properly designed, it can be very inconvenient. At times, for example, water from the washing area could run over to the fuel wood and dampen it. Sometimes, inadequate drainage makes the cooking conditions damp and unhygienic.

Waste management: Leftovers from plates, wasted food, vegetable waste, etc., also require a place to store before they are disposed. Compost bins are a good option. Space permitting, the best option is converting the kitchen waste, biochar and ash into biochar compost. Using biochar compost reduces emissions from the kitchen waste too[10].

Serving area: In institutional kitchens, the place where food is served should be close to the kitchen. If it is not, hot food could get cold by the time it is served, and transportation of the food gets quite inconvenient. Trolleys are a possible solution, but they are rarely available.

A place to wash hands: Washbasins should be located close to the place where the food is served.

Drinking water: Drinking water should be made available close to the place where the food is served.

---

[10] Biochar compost: Biochar (charcoal) has a very large surface area, facilitating microbes for converting the organic matter into compost and absorbs the foul emissions. For more details see: http://biocharindia.com

Bulk fuel wood storage: Fuel wood is usually bought in weekly or monthly basis. There is a need for a waterproof place to store it, ideally on an elevated platform.

Place to prepare fuel wood: The fuel wood that is bought needs to be cut into small pieces, so it can be conveniently fed into stoves. Sometimes this work is done close to the kitchens. The wood, however, also needs to be dried and so an open, vacant space is required additionally. Another point of consideration is the noise made when wood is chopped. This makes it important for the area to be sufficiently far from certain parts of the institution (e.g., as in schools, hospitals etc.)

Collection and storage of ash and charcoal: Ash and charcoal are used for various purposes, such as cleaning utensils. In addition, when mixed with food waste, they make good biochar compost. A place close to the kitchen should be used for this purpose.

Access to the kitchen: Bullock carts / auto trolleys / tractors / other vehicles that bring the fuel wood and supplies to the kitchen should have a good access path to reach it.

Aspect: Sunlight is a very good source and can be used effectively in the preparation of food. An appropriately located kitchen should be able to harness the sun for the greatest part of the day possible. In parts of India, the Southeast corner is the best place for locating of Kitchens. In southern latitudes, the Northeast corner would be the best choice. There are obvious advantages that natural light brings in, when cooking is being done during the day. One can conveniently dry certain food items (such as chillies, pulses, flour, etc.) in the southern courtyard. This is usually done to keep them fresh and free of insects. Traditional architecture and designs consider these factors.

Prevailing wind directions: Say in large parts of India, the prevailing winds blow mainly from the southwest. Locating the stove (facing east) in a southeast corner in the kitchen helps avoid inhalation of smoke. Besides, the wind will be in the same direction as the stove draft, supporting the combustion and improving the stove's efficiency.

Best is the enemy of the Good

Picture 76 Condition of a Government school kitchen

Honour of stove

Nowadays a stove is not honoured as much as other devices in the house. There is a chance that guests visit the toilets but not the kitchens. The host rarely shows the kitchen to the guests. It is common to find kitchens which are unhygienic and untidy. Since it is mostly women who engage in cooking, they have to bear a brunt of these conditions. Cooking three-course meals (as is typically done in India, with *rotis*, rice, curry, pulses, etc.) is not an easy job at all. It becomes even more difficult − at times hazardous − if they do not have access to a good stove.

In parts of India, the stove is believed to embody Goddess *Laxmi* (Wealth) − the Goddess of money and prosperity. This kind of respect means that traditional stoves - made up of clay, dung and bricks - are regularly maintained. Cleaning around the stove and maintaining it is of utmost importance before a

Picture 77 A stove is considered as an important thing in a family

festival. In groups, women go to termite mounds and bring back the mud from them. It is mixed with cow dung and ash to form a paste that is used for plastering stoves. This improves the stoves' life and efficiency. Ash is a good refractory material. When cow dung burns in the stove and turns into ash, it makes for a great insulation material. As straw and other biomass particles in cow-dung paste burn in a stove, they increase the porosity of the material the stove has been plastered with, improving its heat resistance and insulation.

Further, women decorate their stoves with *'Rangolis'*, or design patterns made by applying colour-powders over the mud-dung-ash plaster. This practice illustrates the value accorded to the stove by traditional customs and lifestyles. People believe that the state of

Picture 78 Mud from termite mounds is used for maintaining traditional stoves

the stove reflects whether a household is happy / prosperous, or not. For example, when a cat or dog is found sleeping in a stove, it must belong to a poor household, for it means that this family has not cooked any food for some time.

Stovers

'Stovers' are people who engage with stoves. Currently, stovers come from different walks of life. Very few of them hold any formal qualifications. Some have started their involvement at a rather late stage in their lives.

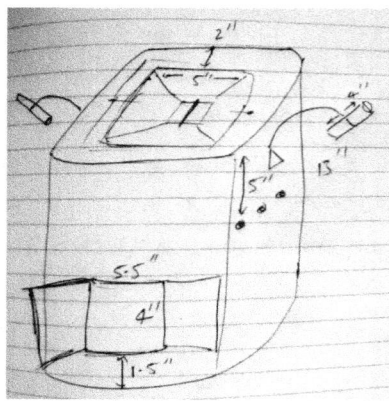

Picture 79 Designing stoves begins with drawings

Stove-related solutions will emerge faster if the scientific community spends more time and resources on design and development. Testing stoves is costly and time consuming. There is a great need for more standard stove testing labs equipped to develop and produce improved designs, and testing them extensively.

Designing a stove for use with a variety of biomass fuels gives much more freedom to the user than one exclusive to a particular fuel type

Designing stoves

Everyone can design stoves. It is the easiest task. One needs to understand that there are mainly three air-streams (Primary, Side and Secondary) to work with. There is a choice between using homogenous or heterogeneous material for construction. It is rare but some schoolchildren are known to design and demonstrate stoves in their science exhibitions. During their

Picture 80 The designer should involve in making stoves

trekking or outdoor adventure trips, people make stoves using cool-drink cans and various other materials. Scientists associated with institutions / universities design stoves using modern equipment and calculated modelling. Organizations working on biomass energy are also contributing to the range of available stove designs. Almost every woman is a stove designer; they design their own stove as per their experience, needs and local trends as they are the ones who cook in most families.

Picture 81 Stoves can be produced using reused / recycled material

A stove designer's mind is never idle. Stoves are ubiquitous; wherever you go, there is a chance to observe them. To a designer, a dustbin in a corner could appear like a stove, awaiting some modifications. Shops selling scrap and kitchen items are ideal for finding design material. Used computer parts (CPUs), flower pots, empty oil cans, used tin cans of various sizes and many other cheap, commonly found things could be seen as design elements. Besides, stoves made of fresh material would be much more energy-intensive (as the embodied energy is high).

While designing stoves, one should not initially aim for the "perfect stove" or the "best stove". Best is the enemy of good. In society, only good things prevail on a large scale, even though everyone wants the best and perfect things. The 'best' is most often costly / inaccessible / not easily replicable / difficult to use / etc.

The stove designer should consider his initial designs to be prototypes. Every prototype stove need not be a successful product. Moreover, every design is a new learning.

Improving existing stoves is the simplest and best way forward towards the 'good stove'. The cost of improved stoves remains low. Local skills and materials can be effectively used for their production and dissemination. They are the easiest for communities to adopt since they are already used to similar designs, albeit less efficient and convenient.

Some of the simple amendments that can improve traditional stoves are:

- Introducing a grate and a primary air source;
- Introducing side / secondary air inlets;
- Introducing lightweight material in the body of the stove;
- Reducing the stove's surface area.

Whenever I visit a new place to develop / disseminate 'good stoves', my biggest priority is trying to improve upon those currently being used.

Comparing stoves with a home

Explaining the principles of combustion to a user is a challenge. In this regard, the analogy of a home can be used. A three-stone stove can be compared to an open shed, which is not convenient and highly insecure as a home. The air enters from all sides without any control. For the same reason, a three-stone stove is the least efficient. The air enters from all sides and it is difficult to cook on. An annular stove[11] is like a shed having 3 to 4 feet high walls covering three sides. The air flow can be excess and it is difficult to cook on in the open. A circular stove with an opening to feed fuel wood, is like a home with only one door. The mixing of air is limited as it can only enter from one side.

A perfect stove with primary, side and secondary air flows is like a home with windows, doors and ventilators with good air circulation. The Magh 3G stove is a good example. A taller stove with a chimney is like a home having high walls; people living in one of these experience good convection currents even without a fan / air conditioner.

---

[11] Annular is nearly circular from top view, with an opening for fuel feed.

Testing stoves

Although there are various methods of testing stoves, the most important testers are the end users. Testing the stoves involves the following of standard procedures. It also requires experience and patience on the part of the tester, a factor crucial to successful testing and designing.

If the tests are conducted during early mornings / late evenings, the air temperature changes very fast. The stove and the pot could be under the warming / chilling effect. Usually,

Picture 82 Testing stoves requires patience and diligence

the places selected for testing are not completely closed, so the local weather and environmental conditions could be influencing factors. Therefore, testing should be avoided during these two points of time during the day.

There are basically three types of tests apart from emission tests:

-   The Water Boiling Test (WBT)
-   Controlled Cooking Test (CCT), and
-   Kitchen Performance Test (KPT)[xvi].

There are several variations of these tests. Their standardisation is still going on. In India, the "Bureau of Indian standards (BIS)" methodology is followed in order to test a stove for approval. Because of variations in the testing methods, thermal efficiency and other parameters vary considerably. (For example, among Magh 3G stoves tested for thermal efficiency by water boiling test: GEO – 28 %, TIDE-27 % and IMMT-24 %[12].

The Water Boiling Test (WBT) is a laboratory test that evaluates stove performance in completing a standard task (boiling and simmering water) in a controlled environment to investigate the heat transfer and combustive efficiency of the stove. They are the easiest, quickest, and cheapest to conduct, but reveal the technical performance of a stove, not necessarily, what it can achieve in real households.

One of the most difficult tasks during WBT testing is weighing the charcoal after the water reaches boiling point. Dousing the flames is convenient when the wood is pushed into the sand.

----

[12] Shown are the average values GEO followed WBT 4.1.2 cold start high power test, TIDE (A Bangalore based organization) followed WBT method and Indian The Institute of Minerals and Materials Technology (IMMT) followed BIS method.

Scrapping all the charcoal from the wood is a difficult task. Some time is lost in weighing the charcoal, the wood and the water in the pot after boiling.

The Controlled Cooking Test (CCT) is a field test that measures stove performance in comparison to traditional cooking methods when a cook prepares a local meal. The CCT is designed to assess stove performance in a controlled setting using local fuels, pots, and practice. It reveals what is possible in households under ideal conditions but not necessarily, what is actually achieved by households during daily use.

The Kitchen Performance Test (KPT) is a field test carried out in real-world settings. It is designed to assess actual impacts on household fuel consumption. KPTs are typically conducted in the course of an actual dissemination effort, with members of the target populations cooking on the test stoves as they normally do. This provides the best indication of how the stove will fare in the real-world scenario.

Human beings are the only species on earth that embraced fire as a source of survival, strength and thriving

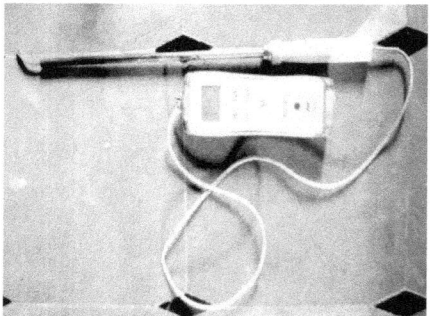

Picture 83 K-type thermometer for measuring temperature of fire

Picture 84 A digital weighing machine

Picture 85 Weighing the charcoal yield

Picture 86 Collection of left over wood and quenching it in sand

Picture 87 Thermometer for measuring the water temperature

Picture 88 Setup for measuring temperature of water

Who can test stoves?

Anyone can test stoves. One needs to have some experience in preparing the kindling wood, lighting stoves, tending to the fire, taming the flames, choosing the wood, feeding it into the stove, observing the fire and nearly 10 different ways of preparing food - roasting, frying, boiling, steaming, simmering and so on.

To carry out WBT tests, one needs to know about boiling and simmering. Biomass stoves are different as compared to other stoves; they are the most difficult to learn and operate. One has to be very careful and have a certain amount of experience. Cooking on biomass stoves can be likened to caring for babies. The best way to experience biomass stoves is to start with three-stone stoves. Anyone who can reach thermal efficiencies of around 30 % with a three-stone stove (in a closed area) is an expert.

Patience and interest are the two most important factors in stove testing. The best way to learn cooking is by observing and learning from the experienced people. Operating a stove professionally for the first time using a manual is difficult. First-time users of a stove should not declare their opinion about a stove in haste.

The highest possible thermal efficiencies are achieved in lab conditions, and only by the experts. Often the highest thermal efficiencies achieved by expert stove testers are reported, while lower efficiencies achieved by operators with less experience are not. Sometimes, stove producers rate their product very high, but they fail to pass through tests specific to certain countries. The highest thermal efficiencies are reported more commonly during the hot-start high power test, than during the cold-start high power test. Most often, the results of the latter are not advertised by stove sellers. It is important to have a standardized system of rating the stoves, such as in terms of certain numbers of "stars", based on their efficiency. For common people it is difficult to understand how so many factors are relevant to a stove performance and evaluation.

Picture 89 A girl testing the prototype Best stove

Promoting stoves

A stove that is tested and endorsed by an organization / agency may not be adopted by a community. The success of a stove is indicated by its adoption by the community that, in turn, affected by the degree of awareness, quality of demonstration and piloting, effectiveness of the feedback

Picture 90 Creating awareness for facilitation of good stoves

mechanism, dissemination strategy, etc. The number of stoves disseminated is the ultimate indicator of whether a stove design is successful.

A stove that has been produced and disseminated on demand by a community would be sustainable, but may or may not be efficient. In many programs, subsidized stoves are distributed or disseminated freely.

> ☐ On promoting stoves – right price of the stove; marketing strategies; developing enterprises and marketing innovatively.

In such cases, the program / scheme could be said to be successful, rather the stove itself. Many stove programs implemented globally have been big failures because they were not sustainable and had to be supported through subsidies / external financing. Besides, the right price is crucial to the stove's uptake by the stakeholder group.

Picture 91 Comparing a traditional three-stone stove and Magh 3G stove - as part of awareness and demonstration

There are many ways of designing stoves using various materials, but the principles of fire and the combustion process are common throughout

Picture 92 Community using different types of stoves

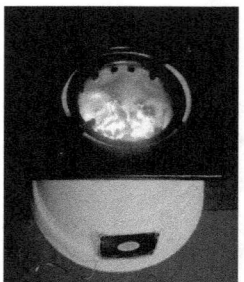

Picture 93 A flower pot as stove

Picture 94 A Computer Processing Unit (CPU) as stove

Picture 95 Twister stove

Picture 96 Avan series best stove made with clay

Picture 97 Good Stove made with bricks and clay

Picture 98 Avani charcoal stove made with used oil tin can

Right price

Stoves themselves are not new to a given community. Even without a dissemination project / program, stoves have always been there and been used. In more than 50 % of the cases, the community spends almost nothing on constructing and installing traditional stoves. They use locally available material such as stones, bricks, clay, dung, sand, ash, etc. Most often, families construct their stoves themselves, using their skills, experience and observations of stoves being used in their region. They do not follow scientific principles much.

Sometimes people buy stoves from their local market; their performance is close to the traditional stoves being used within the community. Such stoves are typically priced between 1 and 3 USD. They are made of used iron sheets, cement or pottery material. These stoves have a slot to feed fuel, a fire chamber and a pot rest. Invariably, a pipe is used for blowing in air during cooking.

We need to understand very well what makes a community willing and ready to adopt 'good stoves'. Ideally, the price of improved stoves should be close to the traditional stoves available in the market. The community is also ready to pay two to three times the cost of traditional stoves, in exchange for increased efficiency and utility aspects. At these prices, 'good stoves' can be facilitated / disseminated without subsidies / schemes

On the other hand, people often understand that good stoves are more efficient and are of greater utility, but still not ready to pay the higher price. Sometimes, they are unwilling to pay it even when they can afford it, since traditional stoves cost them next to nothing. Many a time, the cost of an efficient stove is less than the cost of food cooked for the whole family / the amount spent on doctors and medicines for common ailments / the amount spent on liquor on a special occasion, etc.

If a traditional stove offers hot water as a byproduct, it is preferred over improved ones. If a traditional stove lets you load two pots simultaneously with one common fuel feed source, it becomes very attractive to the user

Other reasons why communities might not adopt efficient stoves (even when they are aware of them) could be lack of knowledge about usage (as in the case of TLUD stoves), or the belief that traditional stoves are better.

Picture 99 A tribal woman cooking on a three-stone stove in an urban slum

Marketing strategies

While setting the price, one should know that a low price is important for wider adoption. Barter trade could also be built into the marketing plan. Even when there are savings in terms of opportunity cost, reduced fuel load and improved air quality, the economically disadvantaged still find it hard to buy a stove that is beyond their operational budgets. There are two options to enable them to access good stoves: reduced costs and/or contextual psychological marketing.

Bringing down the cost, would most probably need subsidies or the use of local labour and raw material in construction. This is a relatively straightforward strategy and probably quite adaptable. Contextual psychological marketing is based on understanding the society and their culture of doing things. For example, in a place like India where there is an established bargain culture, it is possible to price and sell a product by convincing the buyer that the product is of very high value and is coming to them at a bargain price. While it is not possible to set a fixed price for the product, it is possible to sell it at variable prices across households, based on their ability to pay. It works by quoting to them a higher price and then reducing it to the expected price and / or convincing them that the offered price is a special one, reduced especially for them.

Enterprises

Stoves are being made and sold globally. Most of the stoves available in the markets are low-cost and inefficient.

Good stoves often cost more than the stoves currently in use. Traditional artisans or local industries involved in stove production need to be trained in good stoves. Stoves promoted

Picture 100 Traditional artisans trained for production of efficient stoves

through subsidies / government schemes are often costly, one of the many reasons why there is no real demand for them after the scheme is over.

Awareness creation and demonstration are essential to introducing good stoves in a particular area. There are many reasons why few people get into this business. A stove is usually a low-value product that is not sold frequently. The profit margins are low, while facilitation / marketing costs are very high.

Picture 101 Creating awareness to community

Expensive 'good stoves' are often sold at subsidized prices under government projects / schemes. It is important to design and promote good stoves priced close to the stoves currently in the market. It is important to encourage local enterprises, use the skills of local artisans, and encourage local industries, rather than import. Also important are organizations and agencies that can play the facilitating role in the dissemination of good stoves. Stove designs should be declared as "Open Knowledge", so they are accessible to all. If needed, the designers should be compensated by organizations / agencies working on promoting stoves. A wider adoption of good stoves leads to widely shared benefits, some of which are even of global significance.

Picture 102 Awareness on stoves in public places using banners

Enterprises play a big role in making a product sustainable. For the continuous development, improvement and promotion of stoves, there is a need to develop systems. Naturally, there are some adoption-related risks, which, intermediary institutions should try to absorb. These intermediaries could be cooperatives, women's groups, NGOs, technical institutes, government, private companies, etc.

Facilitation of the development and adoption of stoves could follow two basic models. One model involves the distribution of stoves based at one central place, where stakeholders could come and buy the stoves (convergence).

Within the centralized distribution model, a collaborative model between the user and the carbon market supporter could be evolved. The community should be able to buy stoves at a subsidy, and upon returning the stove after use (when the life cycle of the stove is complete), say in one year's time, they would get their money back. Such a scheme would make it possible to monitor the stoves' performance, and meet the goal of promoting their use among more and more users.

Under another model, stoves could be distributed to satellite locations and stoves could be supplied to stakeholders' right at their doorstep (dispersion). The most appropriate strategy to scale up a stove-based enterprise would be a combination of these two models.

Stoves for chicken!

Barter trade has been practiced since long before money became the dominant currency. Most rural homes – even if they do not have cash – have other goods that can be used as currency. Schemes could be introduced that make it possible to pay for a stove with a live chicken. The option to barter will improve the community's purchasing power with respect to stoves.

Share knowledge in such a way that it gives freedom to the person receiving it

Stoves as part of education

Although cooking food is a crucial part of human life, the principles of fire, cooking, etc., are rarely taught as part of formal education. It is common for parents to keep young children away from fire, warning them of its dangers. The Bunsen burner and the hot plate are some of the few heat-based devices that

Picture 103 Involving youth for understanding stoves

children are exposed to in some schools as part of science lab experiments. This should change. Schools should make sure that the children have some basic training in operating stoves, as well as some knowledge of efficient stoves.

Children in rural areas are much more exposed to cooking stoves and fire than urban children. Traditional biomass cooking stoves are commonly used in rural areas. Occasionally, parents ask their children to take care of the stove while food is being cooked, when they have to leave the kitchen to tend to something else. Some children are also engaged in cooking routinely. Unfortunately, this also means that rural children are highly exposed to indoor air pollution when their parents use less efficient stoves in the kitchen.

Usually, children first engage with stoves when they assist parents in cooking. They tend to the flame when parents have to attend to some other work. As they grow up (and especially during the peak agricultural season), many children do most of the cooking in their household. In rural government schools, when I ask children between grades 5th and 10th if they have any experience of operating biomass stoves, most boys and girls say yes. Despite that, their curricula and books hardly discuss principles of stoves or precautions necessary while cooking. A stove is also a means to educate children about pollution (type of emissions), heat and thermodynamics, principles of combustion, health, economy, biochar and ash as byproducts, biomass conservation, biomass for energy, environment, greenhouse gases, global warming, climate change, etc.

A stove is the simplest product and can be produced locally.

Media

Access to diverse sources of media makes it is easy to disseminate technologies. Knowledge can be shared rather fast through the web, blogs, photos, presentations, podcasts, slidecast, videos, e-books, etc. These resources can reach and benefit millions globally. There is also a great need to develop centres for creating awareness, and trainings.

Those who are just learning to design stoves can often get a good idea about a stove through photographs posted on websites. Most often, however, the photographs best highlighting the good performance of a stove are shared widely, and others showing their limitations are not. Unedited videos and pictures are sometimes a better resource to understand a stove, for they capture both its positive and negative aspects.

An analysis of the visitor-profile of stove-related websites gives a fairly good idea of who all are interested in stoves. The highest number of visitors from a country or region need not indicate that people from there are in great need of biomass stoves. Their interest could be based on various other reasons. It is important to note that internet access and the human population are not spread evenly across the earth.

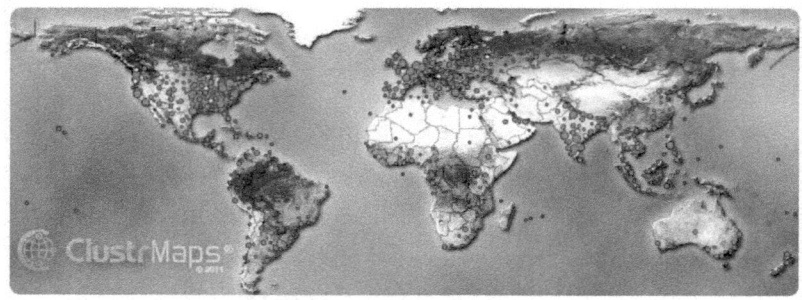

	United States (US)	4,260
	India (IN)	1,918
	Canada (CA)	534
	United Kingdom (GB)	467
	Germany (DE)	423
	France (FR)	391
	Australia (AU)	297
	Russian Federation (RU)	258
	Philippines (PH)	240

Picture 104 Country Totals of the website http://goodstove.com - from 23 May 2010 to 11 April 2012.

The local press plays an important role. Articles on "Good Stoves" draw a huge response from a diversity of people - women, youth, farmers, teachers, institutional heads, government officials, local enterprises, small and large-scale industries, etc. Usually, their interest lies in adopting efficient stoves for their domestic, institutional, enterprise or industrial needs, redesigning their existing stoves, or in willingness to help create awareness.

A diversity of people are showing interest in stoves – traditional artisans / craftsmen (tinsmiths / masons and so on), youth with a variety of educational backgrounds and skills, farmers, teachers, academicians, various organizations / agencies, scientists, politicians and also retired people.

A young person living and working in a city wrote to me, saying he wanted to present a good stove to his mother who lived in a remote area. She had nearly lost her eyesight and had been ill due to over-exposure to smoke. Another enquirer was a lady had just started an herbal beauty products enterprise. She had been using three-stone stoves and could not cope with the smoke. A poultry-owner wanted a stove to provide warmth to young chicks. An entrepreneur in the catering business wanted an efficient stove because the existing stove was consuming lots of biomass fuel, and emitting much smoke, which had his neighbours complaining. Many hotels / lodges want to switch from LPG stoves to biomass stoves, as the price of LPG has gone up really high. Many people engaged in the textile industry (dyeing, spinning, and rolling) wrote in, saying they wanted efficient, good stoves. An industry that uses rotary kiln wants to have a biomass-based heating system. These issues are immensely important to these people and the solutions are simple, but there are not many entrepreneurs dealing in good stoves.

Picture 105 An entrepreneur using Magh CM stove for dressing the chicken

One billion good stoves

We need leaders, leaders and more leaders. Bottom-up leadership is more important for realizing common goals. For facilitating the adoption of good stoves, a number of community leaders need to be identified and involved in the process. These leaders would come from among people living in the villages where good stoves are to be facilitated.

Picture 106 Creating awareness to men

Their leadership potential could be fully realised by creating within them awareness, sensitivity and motivation for facilitating good stoves. They should be encouraged to be part of mitigating local and global problems and to be the "**Earth Leaders**[13]". These leaders are willing to share their experiences with the use of good stoves with

---

[13] Earth Leaders – Anyone could be an earth leader, contributing to the well-being of the earth. These leaders set goals larger than life and achieve them. For example, a farmer in a village facilitating 100 good stoves is an earth leader. A head of a state cannot be an earth leader by declaring the same, but one need to set and achieve common goals as per their position and possibility. Others can also declare someone known to them contributing to sustainability of earth as an earth leader. This is a voluntary declaration. Earth Leaders evolve with the GEO Spirit i.e., Spirit – Knowledge – Action.

other community members. They also create awareness widely in their communities and demonstrate the stoves on demand.

The community leaders work part-time. They are a diverse group – youth, farmers, women leaders, shopkeepers, businesspersons, community musicians, Cable Television Operators, political representatives, officials, teachers, bureaucrats, representatives of various organisations, etc. Some of them coordinate with other community leaders and form larger networks covering large areas. They have very high levels of inspiration; they value their goals as much as their lives.

A leader is not a leader, and everyone is a great leader

Picture 107 Stove designs based on principles for creating efficient stoves

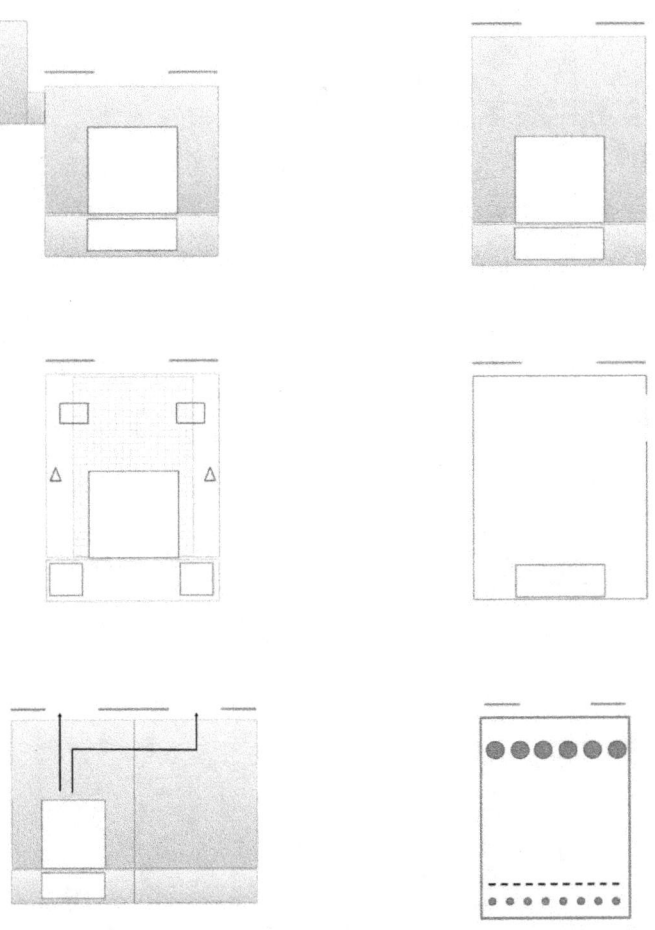

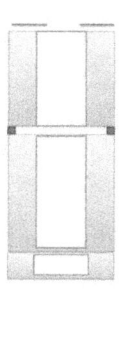

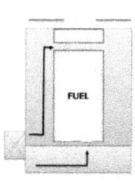

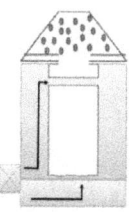

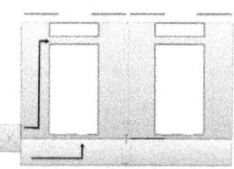

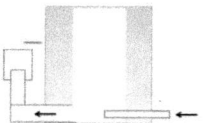

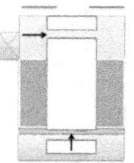

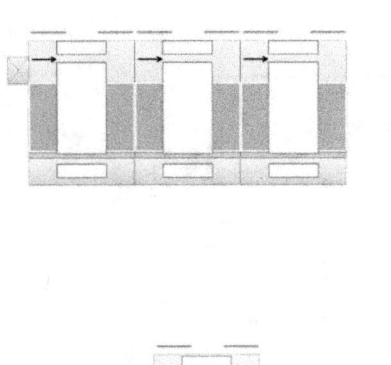

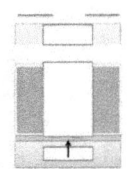

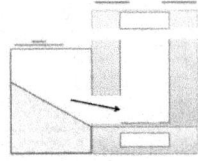

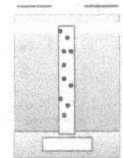

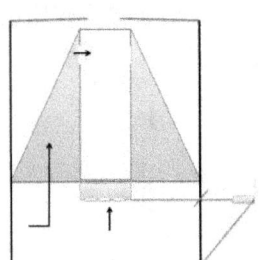

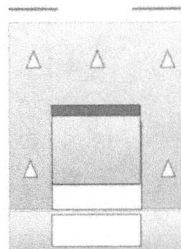

Cycle for sustainable stove development and adoption

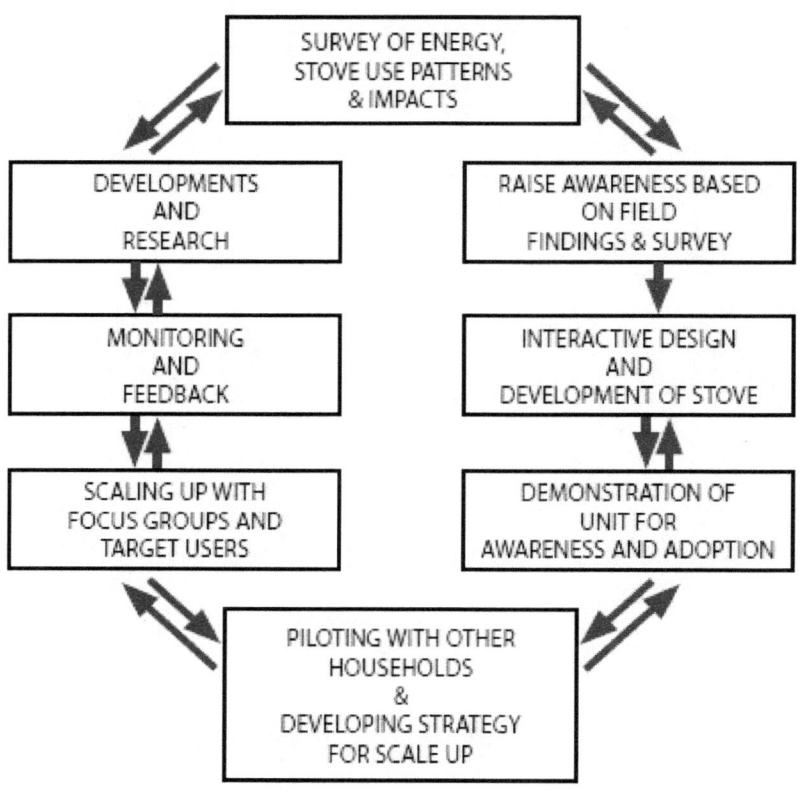

Notes

Volatile organic compound, "all organic compounds of anthropogenic nature, other than methane, that are capable of producing photochemical oxidants by reactions with nitrogen oxides in the presence of sunlight" (McConville, 1997). Volatile organic compounds (VOCs) have a high vapor pressure at ordinary, room-temperature conditions. Their high vapor pressure results from a low boiling point, which causes large numbers of molecules to evaporate or sublimate from the liquid or solid form of the compound and enter the surrounding air.

Polycyclic Aromatic Hydrocarbon (PAH), tar, condensable organic compounds, soot, carbonaceous particles, produced from gaseous fuel or from volatilized solid or liquid fuel components during combustion.

Soot is a general term that refers to impure carbon particles resulting from the incomplete combustion of a hydrocarbon.

Particulates also known as particulate matter (PM), suspended particulate matter (SPM), fine particles, and soot – are tiny subdivisions of solid matter suspended in a gas or liquid. Particles, released when fuels are incompletely burned, can lodge in the lungs and irritate or damage lung tissue.

Health Effects of Combustion Products[xvii]: Carbon monoxide is a colourless, odourless gas that interferes with the delivery of oxygen throughout the body. At high concentrations, it can cause a range of ailments, from headaches, dizziness, weakness, nausea, confusion and disorientation, fatigue in healthy people and episodes of increased chest pain in people suffering from chronic heart diseases. Symptoms of carbon monoxide poisoning are sometimes confused with those of flu or food poisoning. Fetuses, infants, elderly people, and people with anemia or with a history of heart or respiratory disease can be especially sensitive to carbon monoxide exposures.

Nitrogen dioxide is a reddish brown, gas with an irritating odour. It irritates the mucous membranes in the eye, nose, and throat and causes shortness of breath after exposure to high concentrations. There is evidence that high concentrations or continued exposure to low levels of nitrogen dioxide increases the risk of respiratory infection. There is also evidence from animal studies that repeated exposures to elevated nitrogen dioxide levels may lead, or contribute, to the development of lung diseases such as emphysema. People at particular risk from exposure to nitrogen dioxide include children and individuals with asthma and other respiratory diseases.

Acknowledgements

I express my deep sense of gratitude to Mr. Frank van Steenbergen, for his ever support and encouragement in bringing up this book. I am also thankful to Mr. Abraham Abhishek for editing the text and making it more legible and Ms. Linda Navis for constantly guiding me in composing this book excellently. I thank all the other family members of MetaMeta, Netherlands for their support.

I am very much thankful to the stovers Mr. Tom Miles, BioEnergy and Biochar Lists, USA and Mr. Crispin Pemberton-Pigott, Swaziland, Southern Africa, for their encouragement and inspiration. I express thanks to the stovers community, the people who produced stoves designed by me, community using the stoves, earth leaders who supported in facilitation of stoves and all the secondary stakeholders involved in dissemination of stoves in parts of India and abroad.

I would like to thank my sons Avan and Magh, for all their love and support during my last eight years journey into stoves. I also thank my wife and parents for their constant support and giving me freedom. I express my gratitude to my mentors, friends and well wishers for their support in my endeavours.

I express my love and thanks to those millions of women and children whose suffering has sensitized me, and I got an opportunity to work on stoves and mitigate their hardships and address other co-benefits.

I acknowledge and express thanks to my organization 'Geoecology Energy Organisation [GEO]', Hyderabad, India, that has given me space to think, understand, design and innovate on Good Stoves and Biocharculture. Thanks are also to GoodPlanet.org, France for their support to continue my work on Good Stoves. I am very much thankful to MetaMeta, Netherlands for all the support in publishing this book.

27th September 2012   Dr. N. Sai Bhaskar Reddy
Hyderabad, India   saibhaskarnakka@gmail.com

ADOPT GOOD STOVES FOR THE ENVIRONMENT AND

WELL BEING OF ALL

References

Facilitating development and adoption of good stoves at household and community level, Sai Bhaskar Reddy N.: Book - 'At the Bottom of the Energy Ladder', Ed. Peter Kuria, published by Into, Finland. Pg. 52-58, 2011

Magh and Avan series - Designs by Dr. Reddy (India), Sai Bhaskar Reddy N., Book- 'Micro-gasification: Cooking with gas from biomass', Author: Christa Roth, published by GIZ, Pg. 59-60, 2011

http://goodstove.com

http://www.who.int/mediacentre/factsheets/fs292/en/

http://www.sciencedirect.com/science/article/pii/S0973082610000219

http://biocharindia.com

http://e-ruralstoves.blogspot.com/

http://bioenergylists.org

http://www.pciaonline.org/

http://www.hedon.info/

http://cleancookstoves.org/

http://www.winrock.org/

http://www.giz.de/en/home.html

http://www.arecop.org/

http://www.aprovecho.org/

http://biocharindia.com

http://www.epa.gov/ttnchie1/ap42/ch01/final/c01s10.pdf

http://www.pciaonline.org/testing

http://www.epa.gov/iaq/combust.html

http://bioenergylists.org

http://www.epa.gov/ttnchie1/ap42/ch01/final/c01s10.pdf

http://en.wikipedia.org/wiki/Soot

http://en.wikipedia.org/wiki/Particulate_matter

About the author: This is the Stoves Man, a scientist committed to environment, who designed more than 50 low cost efficient good stoves. He has experience in working on Climate Change, Biodiversity, Environment, Natural Resources and Sustainable Livelihoods aspects in parts of India.

MetaMeta Research & Management
Paardskerkhofweg 14, 5223 AJ 's-Hertogenbosch
The Netherlands

www.ingramcontent.com/pod-product-compliance
Lightning Source LLC
Chambersburg PA
CBHW070522210526

45169CB00027B/288